Geothermal Energy

Geothermal Energy

Utilization, Technology and Financing

Kriti Yadav
Anirbid Sircar
Apurwa Yadav

CRC Press
Taylor & Francis Group
Boca Raton London New York

CRC Press is an imprint of the
Taylor & Francis Group, an **informa** business

First edition published 2022
by CRC Press
6000 Broken Sound Parkway NW, Suite 300, Boca Raton, FL 33487-2742

and by CRC Press
2 Park Square, Milton Park, Abingdon, Oxon, OX14 4RN

CRC Press is an imprint of Taylor & Francis Group, LLC

ISBN: 978-1-03-206947-0 (hbk)
ISBN: 978-1-03-206948-7 (pbk)
ISBN: 978-1-00-320467-1 (ebk)

DOI: 10.1201/9781003204671

Typeset in Palatino LT Std
by Newgen Publishing UK

This book is dedicated to our family,

friends, relatives, and colleagues.

The authors also dedicate the book to the

"Geothermal Fraternity".

Contents

Figures

Tables

Foreword

Dear Reader,

Our objective with *Geothermal Energy: Utilization, Technology, and Financing* is to provide a detailed introduction to selected fields of geothermal energy. As geothermal energy is growing very fast, it is difficult to understand the dynamic nature of geothermal sources. When it comes to understanding of geothermal energy, there are limited papers, articles, and books available in the market for the researchers in this field. The aim of this book is to provide information about the geothermal energy sector in terms of not only technology but also financing and its utilization with time.

The authors have decided to publish this book with CRC Press. All the rights of this book will be reserved with CRC Press.

Authors,

Dr. Kriti Yadav, Dr. Anirbid Sircar, and Mr. Apurwa Yadav

Preface

Geothermal energy is growing rapidly worldwide. However, the exploration and exploitation of geothermal energy are still in the nascent and development phase in many countries. Our research on the web and our experience in the exploration and exploitation of geothermal energy made us realize that there is need for a book on geothermal energy with its technical, financial, and environmental aspects. The idea behind the book crystallized in such a way that it could help students, researchers, and professionals in understanding the various aspects and challenges faced by the sector. Papers and references in the public domain, experience in the field, discussion with co-workers, attending and hosting international conferences, visit to Kenya helped to develop the contents of the book, along with geothermal training in Iceland which was conducted by UNESCO and GRO-GTP. The book comprises seven chapters.

The first chapter of the book is dedicated to an introduction to geothermal energy resources. This chapter reviews the basic concepts and classification of geothermal resources and potentials along with the thermal properties of rocks responsible for geothermal potentials. This chapter also discusses the worldwide status of geothermal energy in terms of its utilization and power generation.

The second chapter defines enhanced geothermal systems with subsurface designing, drilling technology, power production, etc. It also describes the environmental attributes and feasibility issues faced in these systems. The third chapter reviews various direct applications in the field of geothermal energy. Applications like sericulture, aquaculture, and honey processing are the emerging direct applications in this sector. This chapter reviews the methods pertaining to sericulture, aquaculture, and honey processing and how geothermal water is used for their processing purposes. The fourth chapter is dedicated to indirect applications of geothermal energy and specifically to milk pasteurization, cloth and food drying, desalination, along with space heating and cooling processes. This chapter helps to understand how geothermal water contributes in these sectors and reduces the heating cost of water required for their processing. The fifth chapter of this book describes the application of the most important technique, which is the need of the hour: machine learning in the geothermal sector. This chapter gives an account of the utilization of machine learning in characterization of the subsurface, drilling, reservoir modelling, and power generation. It describes how with the application of this technique, the issues and challenges in the geothermal sector have been reduced. Chapter 6 reveals the financing sector of geothermal energy.

This chapter talks about the incentives and policies for financing, stages of financing, sources of financing, and issues and barriers to financing in the geothermal sector. The last chapter is devoted to the impact and contribution of geothermal energy on climate change and how to combat this with the application of machine learning. This chapter deals with the mitigation process and factors influencing adaptation. It also describes the use of machine learning in climate change mitigation.

Acknowledgement

The authors are very thankful to Pandit Deendayal Energy University for providing the basic facilities and infrastructure, access to library and internet resources, and an opportunity to research in the sector of geothermal energy. We are also thankful to Government of Gujarat for funding research projects in the field of geothermal energy and establishment of a Centre of Excellence in Geothermal Energy (CEGE).

The authors acknowledge with thanks the assistance given by various companies and individuals in providing information, illustrations, and data for this book. During the course of writing this book, the authors have received help, encouragement, and assistance from colleagues, family members, friends, CEGE members, industry experts, etc. We are highly indebted to all of them. The authors would also like to thank the UNESCO GRO-GTP program for giving them an opportunity to understand the geothermal industry in Iceland.

We thank Pandit Deendayal Energy University for help in the preliminary literature survey for writing the book and giving input on various innovative practices in the global geothermal industry. We are also thankful to the students of the School of Petroleum Technology for their inquisitive questions and discussions, which helped us to improve the content of this book. The authors would like to acknowledge Ms. Widya Asoka from Center for Mineral, Coal and Geothermal Resources – Geological Agency Indonesian Ministry of Energy and Mineral Resources for providing the picture for the book cover which is taken from her drone in Iceland.

Our sincere thanks are also due to CRC Press for giving us an opportunity to publish this book under the auspices of their publication house.

Dr. Kriti Yadav
Dr. Anirbid Sircar
Mr. Apurwa Yadav

Authors

Dr. Kriti Yadav

Dr. Kriti Yadav is a project coordinator at Centre of Excellence for Geothermal Energy, Pandit Deendayal Energy University. Dr. Kriti Yadav has worked as an assistant professor at the School of Petroleum Technology, Pandit Deendayal Energy University from January 2017 to June 2017. She has also served at Centre of Excellence for Geothermal Energy as research associate and research scientist. She has Bachelor's and Master's degrees in geology from Patna University. She has a doctoral degree from the School of Petroleum Technology entitled *Exploration of Geothermal Hot Spots in Gujarat, India*. She also earned a six month training program in geothermal utilization from the UNESCO GRO-GTP Program, Reykjavik, Iceland in the year 2021. The areas of research interest of Dr. Kriti are in the fields of exploration, geology, natural gas, and city gas distribution. Dr. Kriti has published several papers in national and international journals along with conferences. She has published four books in the field of geothermal and city gas distribution. She has published several technical design patents. She has also served as reviewer for several reputed journals published by Springer and Elsevier.

Dr. Anirbid Sircar

Dr. Anirbid Sircar is the Dean of Research and Development of Pandit Deendayal Energy University, India. He served the university as a Director of the School of Petroleum Technology, Pandit Deendayal Energy University, and Director General of the mother institute Gujarat Energy Research & Management Institute. Before joining Pandit Deendayal Energy University, Dr. Sircar worked for oil and gas industries such as L&T, Tullow Oil, HPCL, Enpro, and Jubilant Energy.

The research interests of Dr. Sircar are petroleum exploration and geothermal energy along with natural gas and city gas distribution. He is a professor in the School of Petroleum Technology and serving as Head of Centre of Excellence for Geothermal Energy. Dr. Sircar acted as a member of the task force on geothermal energy established by the Ministry of New Renewable Energy, Government of India.

Dr. Sircar obtained his undergraduate degree in Exploration Geophysics from Indian Institute of Technology Kharagpur and MTech in Petroleum Exploration from Indian Institute of Technology (Indian School of Mines) Dhanbad. He received his PhD degree in Tomography from the same institute. He has successfully guided 34 MTech students and 9 PhD students. He has also completed seven funded projects. Dr. Sircar has published 126 papers in national and international journals, has authored four books, and has seven granted patents. He has also been a reviewer for several reputed journals published by Taylor & Francis, Springer, and Elsevier.

Mr. Apurwa Yadav

Apurwa Yadav is a computer engineer from Silver Oak College of Engineering and Technology. Since 2018 he has been working as an independent researcher in the field of artificial intelligence. He has published several research papers in various reputed journals by Springer and Elsevier. In addition, he has also had the honour to present his work on the platform of the 45th Stanford Geothermal Workshop held by Stanford University. His areas of interest include machine learning and deep learning techniques like computer vision, natural language processing, transfer learning, etc. He has worked in the field of data science and as a software engineer.

1

Introduction to Geothermal Energy Resources

1.1 Basic Concepts and Classification of Geothermal Resources

Thermal energy is a type of energy obtained from the earth's internal heat. The energy of the earth's crust is situated at various depths, ranging from the ground to relatively shallow aquifers to deep aquifers. Molten material components called magma are the underground heat source. Geothermal energy is among the most environmentally friendly sources of renewable energy. Therefore, it has gained popularity in several nations. The term "geothermal energy" is derived from the Greek word *geo*, which means "earth," and *thermal*, which means "heat." Renewable energy in its purest form is provided by natural assets such as sunshine, wind, water, and geothermal heat. The energy trapped in the earth is used to generate geothermal energy. The existence of shallow-level geothermal resources can be attributed to five key variables:

1. Invasion of magma from the depths, releasing massive amounts of heat.
2. Due to a thin crust as well as a significant temperature differential, there is a lot of surface heat flux.
3. Ascension of warm groundwater which has flowed through levels of many kilometres.
4. Thermal inundation or shielding of subsurface rocks by thick formations of low-thermal-conductivity formations like shale.
5. Abnormal warming of subsurface rock caused by radioactive decay, possibly aided by geothermal blanketing.

Therefore, the criteria most often used for identifying geothermal resources are dependent mostly on the temperature of geothermal fluids, which operate as the transmitters of energy from deep heated rocks to

DOI: 10.1201/9781003204671-1

the ground. Temperature is used to describe the specific heat of these liquids and give a general estimate of their value. Temperature is proportional to enthalpy. As per parameters based on the energy concentration of the liquids and their prospective modes of use, the assets are divided into three categories: Low, medium, and high enthalpy. A distinction is often drawn among geothermal systems that are characterized by water or liquid and those that are characterized by vapor (White et al., 1971). As observed in India, water vapor is a constant, resistant fluid component in water-dominated networks. There may be some vapor present, usually in the form of distinct bubbles. Such geothermal resources are the most widespread around the globe, with temperatures ranging between 125 and >225 °C. Depending on the temperature and pressure, it can create hot water, water, steam mixes, wet steam, and even dry steam.

1.2 Thermal Properties of Rocks

The ability of rock material to transmit the heat is known as thermal conductivity. Generally, in boreholes, temperature measurement is done along the vertical profiles by providing vertical component of the thermal gradient. In general, the thermal conductivity of volcanic and plutonic rock is isotropic in nature. As a result, heat in such formations flows in the vertically upward direction. The thermal conductivity of sedimentary and metamorphic rocks is highly anisotropic in nature, which gives significant horizontal heat flow. In the case of a horizontal heat gradient, the result can be obtained from laboratory measurement of core rock samples in different directions.

1.3 Geothermal Technology of the World

Electricity produced by heat trapped underground is known as geothermal energy. As this heat extraction is tiny in comparison to the earth's surface temperature, it is considered a viable, renewable energy supply. Geothermal power facilities emit only about 5% of the greenhouse gases emitted by coal-fired power stations in general. Worldwide, total geothermal energy capacity has increased from approximately 10 GW in 2010 to approximately 13.3 GW in 2018. Table 1.1 represents total geothermal power installed capacity in different nations.

TABLE 1.1

Installed Capacity of Geothermal Energy in Various Countries

Country	Installed Capacity (GW) in 2018
United States	2.50
Philippines	1.93
Indonesia	1.95
Mexico	0.9
New Zealand	1.0
Italy	0.8
Iceland	0.7
Turkey	1.3
Kenya	0.6
Japan	0.5

1.3.1 Geothermal Technology

Over the last hundred years, geothermal power production has become centred on regions with abundant hydrothermal reserves. Moreover, with the advancement of modern tools, geothermal energy may now be produced in low-enthalpy areas as well. There are two ways to employ thermal techniques:

1. Geothermal Electricity
2. Geothermal Heating and Cooling

1.3.1.1 Geothermal Electricity

a. Flash Steam Plant: Uses steam and hot water from inside the earth to rotate the turbine and generate electricity. It is better for high-temperature reservoirs. Generally used in the United States.
b. Dry Steam Plant: Uses dry steam produced from the reservoir to rotate the turbine and generate electricity. It is beneficial for reservoirs with high-temperature steam.
c. Binary Power Plant: A heat exchanger is used to heat geothermal liquids, which evaporate at a lower temperature than freshwater. Vapor in the atmosphere is re-pressurized and transferred to the exchanger in a closed system after turning the turbine to produce electricity. Low-enthalpy aquifers benefit from this approach (Yamamoto, 2001).
d. Enhanced Geothermal System (EGS): A thermal resource that has been artificially produced or upgraded. It permits geothermal electricity to be generated almost anywhere with a low or medium temperature.

1.3.1.2 Geothermal Heating and Cooling

a. Extremely low temperature at the limit of the average annual ambient temperature at the location, up to roughly 25°C: This is dependent on steady aquifer and subsurface temperature at shallow depths (about 400 m). Ground source heat modules are generally used to harvest energy from the subsurface and increase the temperature to the desired level for space heating (Lund et al., 2005).
b. Low and moderate temperatures, varying from 25°C to over 100 °C: Heat is extracted from the ground and aquifers at a higher temperature and, in most cases, at a deeper level. The energy from the earth or the aquifer could be used directly if the geothermal water is at a temperature that is consistent with the appropriate temperature for the heating system.

1.3.2 Global Direct Use of Geothermal Energy

a. Around 82 nations use geothermal energy for direct application. Nations such as China, Denmark, Sweden, etc. in the northern hemisphere mainly use geothermal energy for direct application (Fredleifsson, 2001).
b. Geothermal energy is progressively being channelled into space heating systems (over 230 in Europe) and/or larger public buildings (malls, offices).

1.3.3 Stages of Construction and Cost of Setting up a Geothermal Power Plant

Geothermal power expenses per MW vary from 4 to 7 million euros (i.e. from INR 300 million to INR 525 million). Drilling expenses are determined by the following factors:

a. Number of wells to be drilled
b. Depth of the wells drilled

1.4 Geothermal Energy Potential of India

1.4.1 History

Early studies of geothermal energy by the Geological Survey of India (GSI) have suggested that geothermal energy production has potential in India.

TABLE 1.2

Geothermal Provinces Hosting More Than 340 Thermal Springs

Orogenic Geothermal Energy Provinces	Sr. No.	Non-Orogenic Geothermal Energy Provinces
Himalayan	4	Cambay graben
Naga-Lusai, Sikkim, Arunachal, and Bhutan	5	Son-Narmada-Tapi graben
Andaman-Nicobar Island	6	West Coast
	7	Damodar valley
	8	Mahanadi valley
	9	Godavari valley
	10	North Indian Peninsula
	11	East Indian
	12	South Indian

(GSI, 1991, Sircar and Yadav, 2021)

Depending on their prevalence in certain geomorphological settings, hot springs have now been clustered together and labelled as various geothermal regions. There are 12 geothermal provinces hosting more than 340 thermal spring fields in India, as given in Table 1.2.

1.4.2 The Major Geothermal Manifestations of India

The major geothermal manifestations of India are given in Figure 1.1. Eight of them are identified as promising areas for harnessing geothermal power or direct heat applications. The summary of these eight major geothermal manifestations is given in Table 1.3.

1.4.2.1 Puga Geothermal Field, Leh District, Ladakh, J&K

The Puga geothermal field along Puga nala (i.e. a tributary of the Indus River) in the Ladakh region of J&K extends over an area of 7–8 km^2 and is the most promising field, having 120 thermal springs and pools with surface/spring temperature up to 84 °C and hot water cumulative discharge up to 300 l/s. Besides, there are sulphur condensates along joints/cracks, and borax evaporates as loose deposits over the valley floor. Out of 34 boreholes of 28.5–384.7 m depth drilled by GSI, 17 boreholes have hot water–steam mixed flow in artesian conditions. The geochemistry of the thermal water shows high concentrations of chloride, boron, sodium, potassium, and other cations along with high total dissolved solids (TDS) (1900–2500 ppm), clearly indicating high temperature at depth. Based on silica geothermometry, the estimated temperature of the reservoir at Puga is 180–225 °C (Chandrasekhram, 2015).

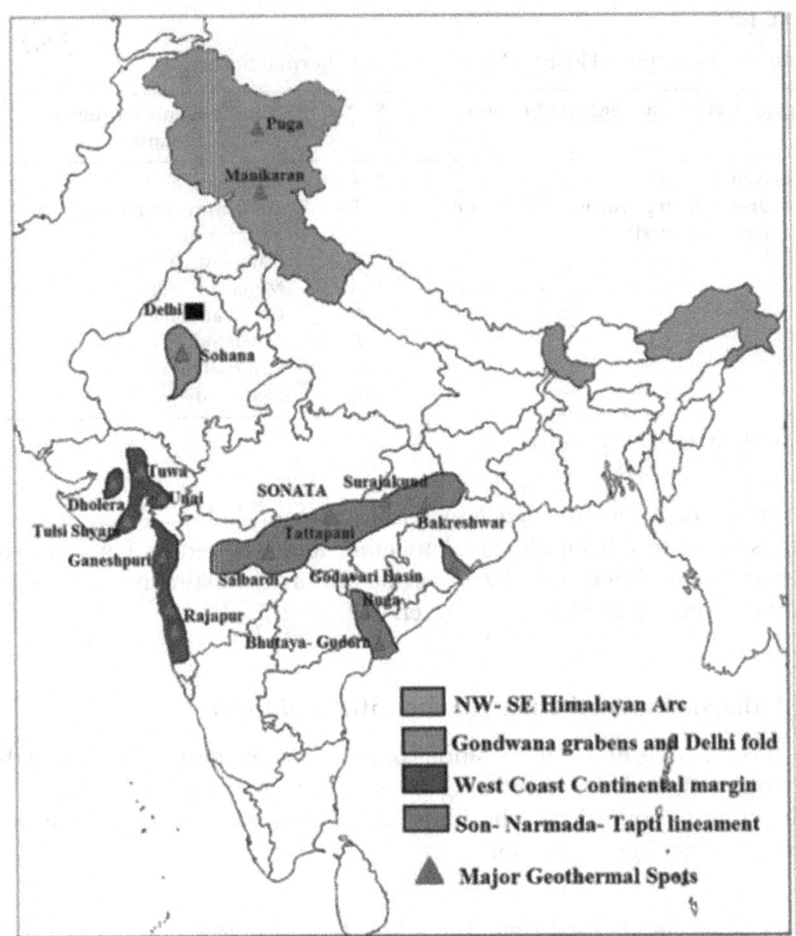

FIGURE 1.1
Major geothermal manifestations of India.

The Regional Research Laboratory, Jammu has installed a greenhouse hut using geothermal water for mushroom cultivation and preservation at Puga, Ladakh, J&K. The geochemical data along with the integration of geophysical and geological data indicates that a 3 MW geothermal power plant can be installed at Puga for the benefit of local villages (Manikandan, 2015).

1.4.2.2 Chumathang Geothermal Field, Leh District, Ladakh, J&K

At Chumathang (4200 m height), many hot springs with surface temperature ranging up to 90 °C spread over a distance of about 0.7 km along the

TABLE 1.3

Major Geothermal Manifestations in India

Himalayan Region
Puga Geothermal Field, Leh district, Ladakh, J&K
Chumathang Geothermal Field, Leh district, Ladakh, J&K
Manikaran Geothermal Field, Kulu district, Himachal Pradesh
Tapoban Geothermal Field, Chamoli district, Uttarakhand
Peninsular Region
Tattapani Geothermal Field, Balrampur district, Chhattisgarh
Manuguru Geothermal Field, Bhadradri district, Telangana
West Coast Geothermal Field, Maharashtra
Surajkund and Tantloi – Bakreshwar, Jharkhand, and W.B.

(Yadav and Sircar, 2021)

right bank of the Indus River. The isotopic study conducted at this location does not reflect any major shift, which indicates that the thermal waters are meteoric in origin. The thermal waters do not show high concentrations of boron, chloride, and other ions and have low TDS as compared with Puga (1000–1250). Based on silica geothermometry, the reservoir temperature varies between 165 °C and 180 °C, and there is no evidence of any magmatic contribution at this site. Six boreholes were drilled in Chumathang by GSI and recorded 87 °C to 120 °C subsurface temperature and up to 700 l/min cumulative discharge. Four of these boreholes yielded hot water and steam mixture. At present, most of the boreholes have been choked. Under Indo-Norwegian collaboration, a space heating pilot project is running to warm the Lamying Restaurant, Chumathang.

1.4.2.3 Tatapani Geothermal Field, Balrampur District, Chhattisgarh

The Tatapani geothermal field, along the eastern margin of the Son-Narmada-Tapi (SONATA) rift zone, is one of the most resource-abundant geothermal manifestations in Peninsular India. The area has been extensively studied by GSI to assess the resource potential for development of geothermal energy. For this purpose, 26 exploratory boreholes were drilled during the period from 1980 to 1995 to establish the subsurface geology, extent, and subsurface reservoir condition of geothermal resources and to evaluate resources up to 350 m depth for a pilot power generation plant. GSI established the feasibility of generating 300 KWe of electric power over a period of 20 years and inferred the reservoir potential to be 17 MWe based on various geothermal parameters. Geothermal power potential of 40 MWe by binary cycle method is postulated for the depth range of 1500 to 2000 m.

Under the provisions of a memorandum of understanding (MOU) between GSI and National Thermal Power Corporation Limited (NTPC), a File Sharing Protocol (FSP) item entitled "Exploration for Geothermal Resource Assessment at Tatapani Geothermal Field, Balarampur District, and Chhattisgarh" was undertaken in the year 2014–15 to prepare a Detailed Geological Project Report (DGPR) for geothermal power plant feasibility. Detailed geological, geochemical, isotopic, and resource calculations have been attempted. The area presently has seven springs with temperature ranging from 63.6 °C to 88.7 °C. The study suggested that the temperature in the geothermal reservoir might be between 150 °C and 190 °C. The geothermal geothermometers and thermal logging data indicated that there may be a continuity of temperature >125 °C below 1000 m depth. Therefore, the total available heat in the geothermal system is calculated as 169,072 × 10^7 MJTH over an area of 4.94 × 109 m^3 at a depth of 70 to 2000 m.

1.4.3 Potential Geothermal Resources and Their Prioritization

Based on re-evaluation of data, advancements in technology, changed ground realities, need-based requirements, and interaction with experts, it is proposed that to start with, the geothermal prospects listed in Table 1.4 in order of priority may be taken up for development and demonstration purposes.

The next list consists of some of the prospects that hold promise for both electrical and non-electrical utilization schemes. These are basically of three types:

Group I: Prospects from which power generation may be feasible provided it is consumed locally.

Group II: Prospects from which power generation is possible but not feasible.

Group III: Prospects that may be harnessed for non-electrical applications, particularly for enhancing tourism amenities (Yadav and Sircar, 2021).

Table 1.5 describes Groups I to III.

1.5 Steps to Promote Geothermal Energy in India

Resource Assessment/ Pre-feasibility Study: In the initial stage to promote the confidence of developers and investors, government can set up

TABLE 1.4

Significant Geothermal Hotspots in India That Can Be Taken Up for Field Development in Order of Priority

Name of the Prospect	State	Significant Characteristics	Remarks
Puga–Chumathang	J&K	Most promising geothermal prospects in country with expected resource temperature of >200 °C.	Ideally located in a power-starved region. May be used for heating too. Diesel power generation may be alleviated. Army and local tribal population would be the greatest beneficiaries.
Tattapani	Chhattisgarh	Considered the best geothermal prospect outside Himalayas. Temperature >100 °C at very shallow levels.	NTPC is in the process of developing a technology demonstration project here. The tribal belt would be the beneficiary.
Surajkund	Jharkhand	Surface temperature 88 °C. The temperature of the resource may be 150 °C. Resource is very similar to that at Tattapani.	Both power generation and cooling may be possible. Beneficiary would be the local tribal population.
Bakreshwar	West Bengal	Large thermal anomaly comprising at least seven hot springs with surface temperature of 60–68 °C. Expected resource temperature may be in excess of 120 °C.	Experimental power generation and cooling may be possible.

pilot projects through incentivized schemes along with conducting the resource assessment. Also, the government can sponsor initial resource assessment to be done by private players. These resource assessments can be further used to create a database and allot land/projects to developers. The resource areas/land can be auctioned through competitive bidding by respective state/provincial governments to international/national consortia of geothermal power developers, entrepreneurs, industries, and financial institutions through various models, such as royalties based on thermal equivalency to coal/gas, leaving all technical and financial aspects to the bidders.

Revenue Policy: Financial support measures are designed to attract private sector investment in thermal ventures by improving the value or predictability of the project's income and helping with the project's initial capital requirements:

TABLE 1.5

Classification of Prospects in India for Direct and Indirect Uses

Name of the Prospect	State	Significant Characteristics	Remarks
GROUP I			
Nubra Valley	J&K	Expected resource temperature >150 °C, which is good enough in high-altitude area for power generation	Power generated to be locally consumed. Army and population of Panamik, the base camp of Siachin forces, may benefit.
Tatta	Jharkhand	Expected resource temperature of about >150 °C	Resource temperature may be appropriate for power generation using binary cycle. May be used for cooling too.
Jarom	Jharkhand	Expected resource temperature of about >150 °C	Resource temperature may be appropriate for power generation using binary cycle. May be used for cooling too.
Tural-Rajwadi	Maharashtra	Expected resource temperature of about > 150 °C	Power generated may be locally consumed.
GROUP II			
Tapri	Himanchal Pradesh	Resource temperature of about >150 °C	Surplus hydro power makes geothermal power generation not feasible in the area.
Badrinath	Uttarakhand	Resource temperature may be up to 150 °C	Surplus hydro power makes geothermal power generation not feasible in the area.
GROUP III			
Rajgir	Bihar	Resource temperature may be around 100 °C	May be used for enhancing tourism amenities in the area.
Tapoban	Uttarakhand	Resource temperature may be up to 150 °C	May be used for enhancing tourism amenities in the area around Badrinath.
Manikaran	Himanchal Pradesh	Resource temperature may be up to 150 °C	Maybe used for enhancing religious tourism further in the area.

a. Feed-in tariff rates are determined by the government and give the lowest price, and therefore revenue, that enterprises can anticipate for thermal facility energy production. In nations such as the United States, Germany, the Philippines, and Kenya, they are the most popular type of financial assistance for geothermal energy. Bonus feed-in tariffs are also used by countries like Germany and Turkey to stimulate specific types of power generation or use of domestic manufacturing goods. The feed-in tariff for geothermal power project considering the following assumptions comes out to INR 6.0 to 8.0 per unit.

Project Cost	**INR 300 Million per MW**
PLF	70%
D/E	80:20
Loan Tenure post Certificate of Deposit (COD)	25 years (2 years moratorium and 23 years of repayment)
Interest Rate	9.0%

b. Quantification requirements, like the Renewable Portfolio Regulations, set a goal for renewable energy production over a set period of time. They've been a major proponent of geothermal management in the United States. Because geothermal power plants produce around four times the amount of power compared with other renewable energy sources, such as solar power stations, they may be given equal importance under the Recruitment Process Outsourcing (RPO) duty to compensate for the large original investment.

c. Competitive tenders are really a tender process for the construction and operation of a geothermal power plant of a certain capacity at the electricity price set in the successful bidder's purchase agreement with the power distributor (Indonesia, the United States). These do have the ability to increase competition in the banking sector costs, but they can also result in complicated transactions for plants.

d. Mechanisms for well production: For well exploration and development, government can take the following models:
   i. Production Sharing Model: As per the production sharing model, the revenues from the production are shared between the producer and the user/government.
   ii. Cost Recovery Arrangement: As per the cost recovery model, an operator is allowed to recover all his cost before sharing profit with the user/government.

1.5.1 Monetary Policies

Financial support policies lower the tax rate, boost total profits, and lower operational costs for geothermal ventures by changing fiscal laws.

a. An expenditure tax credit equivalent to a proportion of the geothermal expenditure is provided to the operator. For instance, in the United States, geothermal heat pump equipment installed before the end of 2016 was eligible for a 10% expenditure tax credit. Its effect is contingent on the entity having sufficient income to qualify for the tax cut.
b. In the United States, production tax credits are in the manner of a per-kilowatt-hour tax credit for power generated by geothermal plants.
c. Enhanced Depreciation: Governments, such as those in Mexico, the Philippines, and the United States, grant a term of enhanced depreciation of some or all of a program's investments, based on certain operating standards. As a result, a quick return on capital investment is possible.
d. Lateral Use of Land: Land allotted for setting up geothermal projects can be used for other developmental projects as well, like setting up solar power plants, promoting tourism, and nature and community development. This will not only lead to development of the community nearby but will also lead to additional revenue being generated by the project developer.

1.5.2 Decreasing Investment Funding Costs/Risk Prevention

Loan guarantees and subsidized financing terms are two tools that can help you save money on your funding financing costs. These types of assistance are required when dividend payments are insufficient to pay investors for liabilities they perceive, which could be a problem in the early stages of geothermal development (Mazzucalo, 2018). The Indian government may ask a coalition of enterprises, whether Public Sector Undertakings (PSUs) or foreign firms, with competence in the geothermal and/or power sectors to launch a pilot study.

a. Loan guarantees: In the case of a debtor fail, a government body, multilateral development bank, or public institution offers a lender a guarantee of partial or full debt repayment. This encourages private investment without necessitating or requiring urgent recourse to governmental funds.
b. Public grants (including concessional loans): As in Kenya and Indonesia, a government institution, multilateral development bank,

or public institution gives a grant to help with project expenses or a tax concessions loan with below-market rates and terms. Surface exploration and validation drilling can be funded (or risk shared) by the government if private companies take on the risk of confirmation and commercial drilling.

c. Public security is when the government reimburses a portion of the cost of failed drilling operations. If the exploratory stage of the project doesn't succeed in a predefined level of performance, the refund is given. This was regarded as a critical policy in the growth of Iceland's geothermal industry. As the sector has gained more expertise, it has been used less and less, but there has been minimal imitation in other nations.
d. Insurance Facility by Third Parties: To limit the geotechnical and exploratory risk, insurance coverage is required. This type of insurance boosts the program's credit profile and boosts the vendor's and investor's credibility. Munich Re, for instance, is insuring a set of eight boreholes in a Kenyan region. An exploratory insurance policy protects against inadequate output due to a lack of geothermal resources and is frequently required for geothermal project funding.
e. Capacity Building: Geothermal energy development in India is mostly in the nascent stage and requires technical knowledge and expertise in setting up a project. Capacity building and understanding the latest technology will not only help in faster implementation of the project but will also help in reducing the dependence on developed countries.

1.5.2.1 Challenges

a. Low technological advancement
b. High cost in setting up the project
c. Large financial requirement
d. Resource and other risks
e. Long gestation period
f. Lack of institutional capacity
g. Shortage of professional skills (scientific, technical, commercial, legal)
h. Suboptimal legal and regulatory framework; environmental, social, and cultural impacts

From this detailed study, it is obvious that commercial geothermal projects are not financially viable when the tariffs for other technologies are as low as Rs 2.44 per unit; however, potential Revenue Expenditure (RE) in geothermal energy may be exploited by providing government support for a demonstration project. Once risk assessment has been done on the demonstration project, its techno-commercial viability will be worked out, and policy will be formulated to make geothermal projects in the country commercially viable.

1.6 Zoning of Geothermal Provinces in India

Based on data available from exploration, oil field experience, drilled oil wells, depth prognosis, and conceptual field development, geothermal provinces in India can be classified into four zones (Table 1.6).

1.7 Some Financial Models of Countries Exploiting Geothermal Energy

Some project financing examples of geothermal projects in the United States, the Philippines, and Kenya are described in the following subsections (World Bank, 2020).

1.7.1 United States: Financing of Miravalles III Project

- During 1995, Costa Rica's Autonomous Power Production Act was modified to allow, for the very first time, private participation in the energy industry. The government-owned Instituto Costarricense de Electricidad (ICE) took advantage of the legislation in late 1996 and put Miravalles III up for international bid.
- A multinational partnership of power providers, including Oxbow Power Services of Reno, Nevada, and Marubeni Company of Tokyo, won the contract in the year 1997. To continue funding this $70 million project, the InterAmerican Development Bank (IDB) organized a $49.5 million loan, which was partially guaranteed by other financial institutions.
- Under the terms of the agreement, the partnership will construct the plant, run it for 15 years, distribute the power produced to ICE during that time, and afterwards hand over the power station to ICE at the end of that time.

TABLE 1.6

Geothermal Zones of India

Head	Zone 1	Zone 2	Zone 3	Zone 4
Field	Himalayan Belt 1. Puga 2. Chumatang 3. Nubra 4. Kargil	Oil recovery fields Gujarat Rajamundry, Andhra Pradesh	Known fields 1. Khammam (Telengana) 2. Chinatlpudi (Andhra Pradesh) 3. Tattapani (Chhattisgarh)	Exploration stage 1. Orissa 2. Maharashtra 3. Telengana 4. Arunanchal Pradesh
Resource type	High temperature 250–350 °C	Medium temperature 160–180 °C	Medium temperature 120–180 °C	Medium temperature 100–180 °C
Theoretical potential (MWe)	2500	3000	100–300	Up to 500
Achievable potential (MWe)	250–400	500 (minimum)	75–150	Further investigation needed
Justification of potential	Current exploration data, read in conjunction with UNDP and China geological survey data	A minimum of 1000 abundant oil wells. Recovery factor 0.5 to 3 MWe	Current exploration data, read in conjunction with Australian geothermal developers' survey data	Based on initial exploration models
Field studies	1. Geophysics 2. Geochemistry 3. Deep exploration drills 4. Economic survey	1. Geochemistry 2. Production drills	1. Geophysics 2. Geochemistry 3. Gradient drills	1. Geophysics 2. Geochemistry
Production well drill point located	Yes Point-specific survey needed	Well exists	No Further investigative studies needed	No
Drill rig deployment readiness	No	Well exists Only service rigs needed	No Further investigative studies needed	No
Estimated depth (m)	1500–2200	2500–4000	2200–2700	>2000

(continued)

TABLE 1.6 (Continued)

Geothermal Zones of India

Head	Zone 1	Zone 2	Zone 3	Zone 4
Accessibility to site for rigs and equipment Scale 1–10 1: Easy 10: Difficult	7	1	1	1
Estimated life without reinjection (years)	25 (min)	5–10 (existing well)	15–17	Unknown. However, reservoirs have plenty of natural recharge from sea and western ghats
Major drill rock type	Igneous/Basalt/Hard rock	Sedimentary/Mesozoic	Igneous/Basalt/Hard rock	Igneous/Basalt/Hard rock/Sedimentary
Know drill risks	Estimation of condensable and non-condensable gases not done High sulphur content	Drilled into hydrocarbon reservoirs, risk of wild wells, methane flaming, heavy carbon residue	Further investigative studies needed	No gradient or exploration or slim wells
Auxiliary minerals extractable for industry	Sulphur and borax	Nil	Nil	Further investigation needed
Known production/ injection well operating risks	Heavy calcification of wells, and gas discharges due to shallow seduction zone	Proper residual gas separators and marginal hydrocarbon treatment and separate disposal needed	Further investigative studies needed	High NaCl corrosion risk due to proximity to sea
Plant type	Flash turbine and combined cycle	ORC	ORC	ORC

Transmission status	Lack of proper evacuation facility at Ladakh Single 33 kV line planned and being built up to Chummatang Ideal location for evacuation to central grid in Uttarakhand, Haryana	Well connected by state and central grids Substation within a 3–5 km radius	Well connected by state and central grids Substation within a 3–5 km radius Tattapani substation at 15 km radius	Well connected by state and central grids
Well system overhaul (years)	7 (minimum)	Geothermal production wells needed on completion of well life	10 (maximum)	Further investigation needed
Plant overhaul (years)	7–12	15	15	15
Cost of surface power plant (USD/MW)	1 to 1.5 million	1.8–2 million	1.4–1.9 million	Further investigation needed
Cost of production well drilling casing with drill pad development (USD/well)	3–4 million	0.25–0.5 million for well service	2–3 million	Further investigation needed
Cost of exploration drilling with drill pad development (USD)	4–5 million	Well existing	3–4 million	1–3 million, gradient and slim well only
Estimated number of production well to injection well ratio (P:E ratio)	2:1	1:1	3:1	Further investigation needed
Estimated cost of least environmental damage waste management protocol	1. Direct injection of condensate brine 2. Sulphur and borax treatment with well head inhibitor	Treatment of waste hydrocarbon residue and inject back treated brine Safe disposal of waste hydrocarbon residue	Direct reinjection for all brine	Reinjection after treatment and removal of excess NaCl

(continued)

TABLE 1.6 (Continued)

Geothermal Zones of India

Head	Zone 1	Zone 2	Zone 3	Zone 4
Known plant operating risks	1. Altitude management 2. Employee health management 3. De-icing of SAGS, well head and plant field 4. Variable geochemistry	1. Wild wells 2. Unsteady flow from wells 3. Residual gas and waste hydrocarbon treatment at well head	Further investigation needed	Further investigation needed
Social impact factor	Lowest land/MW ratio of any known renewable energy Socio-economic development Highest direct indirect jobs/MW Guaranteed employment over a 25-year tenure Support winter tourism	Lowest land/MW ratio of any known renewable energy Employment to displaced farm owners Re-utilization of abandoned investments in well infrastructure Revenue as royalties to government PSU	Further investigation needed	Further investigation needed
Current need for renewable electricity in zone	94 MWe needed by 2020 159 MWe needed by 2025 Ladakh hill development road map document Current diesel generators generate at 17 Rs/kWh with 50% subsidy	RPO already fulfilled with name plate solar capacity	RPO already fulfilled with name plate solar and wind capacity Power surplus region	RPO already fulfilled with name plate solar and wind capacity Power surplus region

Policy challenges	Lack of administrative oversight and clarity between LAHDC and PDD/PDC Lack of administrative oversight and clarity on nodal agency to issue environmental clearance Reluctance by nodal agency to support altitude- and location-specific FiT Policy paralysis Geothermal energy not notified part of state RPO	Lack of administrative oversight and clarity with nodal agency on ownership of land Lack of administrative oversight and clarity with nodal agency on ownership transfer of well Lack of central government policy clarity on ownership, use, and legality of non-producing HC wells Geothermal energy not notified part of state RPO	Lack of administrative oversight and clarity on nodal agency to issue environmental clearance Reluctance by nodal agency to support resource-specific FiT Policy paralysis Geothermal energy not notified part of state RPO	Lack of administrative oversight and clarity on nodal agency to issue environmental clearance Reluctance by nodal agency to support altitude- and location-specific FiT Policy paralysis Geothermal energy not notified part of state RPO
Social challenges	Will support major social and economic upliftment Power in winter and winter tourism Reliable base load power for defence forces Eliminate health hazards due to black carbon and diesel burning at high altitudes Diesel can be put to better use in Himalayan belt Helps offset climate change and reduce speed	Employment to displaced farmers and ability to restart family earning Micro grids ranging from 0.5 to 3 MWe to help local community with cheap and reliable power Since wells already exist, location is defined; hence, need does not exist to connect all to central grids Power solutions can be customized to local community-specific needs	Further investigation is needed Proximity to holy places having natural thermal springs (e.g. Bugga, Manuguru, Badrachalam, Tattapani) can pose a challenge	Further investigation is needed

(continued)

TABLE 1.6 (Continued)

Geothermal Zones of India

Head	Zone 1	Zone 2	Zone 3	Zone 4
	of glacier retraction Proximity to holy places having natural geothermal springs (e.g. Manikaran) can pose a challenge			
Economic challenges	Higher Capex due to location altitude and terrain Need support for altitude and location FiT	FiT policy must take into account the higher risk due to shorter life span of well Unexpected gas ejection may further reduce life of well	Need support for low-temperature-specific FiT Higher failed production wells probability	Further investigation is needed
Promotion tariff to start				
Environmental damage factor	1	1	1	1
Scale 1–10 1: Low 10: High				

FiT: Feed in Tariff, HC: hydrocarbon, LAHDC: Ladakh Autonomous Hill Development Council, ORC: Organic Rankine cycle, PDC: Power Development Corporation Limited, PDD: Power Development Department, PSU: Public Sector Undertaking, RPO: Recruitment Process Outsourcing, SAGS: Steam Above Ground System, UNDP: United Nations Development Programme.

- This venture, termed a Build-Own-Transfer (BOT), is the first of its kind in Costa Rica and the very first thermal BOT in Latin America to complete construction. Miravalles III can be cited as evidence of how privatization can operate in the geothermal sector, particularly in terms of funding and administration.

1.7.2 Philippines (Asian Development Bank, 2018): Setting up of Plant by PNOC

- The PNOC Energy Development Corporation (PNOC EDC), a government and regulated company, funded its geothermal operations through World Bank (WB) and Japan Bank for International Cooperation (JBIC) Official Development Assistance (ODA) loans. Via sector loans, the WB assisted the energy industry in financing exploration drilling and demarcation of many locations to help the Philippines realize its geothermal potential. Following the construction and completion of 777 MW of geothermal reservoirs and power stations, WB operation loans were used to fund the development and completion of geothermal fields and power stations.
- The 112.5 MW Tongonan I Thermal Project on the island of Leyte was the very first significant geothermal business venture in the Philippines for PNOC Energy Development Enterprise (PNOC EDC). Under the Eighth Yen Credit Package of the JBIC, previously recognized as the Overseas Economic Cooperation Fund (OECF), funding for the venture was organized in a package that included both geothermal improvement of production (for implementation by PNOC EDC) and power station renovation (for execution by National Power Corporation (NPC)). The grant for steam field development included a JPY 5.8 billion contribution from PNOC EDC. The credit agreement was signed in the year 1980, and the construction was completed in March 1983.
- Around a year later, the JBIC inked a loan agreement for the 112.5 MW Palinpinon I Geothermal Electricity Project as part of the Ninth Yen Credit Program. The PNOC EDC's contribution to steam field development was JPY 10.8 billion. During June 1983, Palinpinon I was approved for access to the funds.

1.7.3 Kenya: Setting up Olkaria III Project

- The Olkaria III geothermal plant is Africa's first privately financed and constructed geothermal plant. A staged development approach, as well as a blend of public and private funding and risk prevention tools, secured the project's feasibility.

- Olkaria III is a 110 MW binary geothermal electricity plant whose reserves were first studied by government before being developed by Ormat Technologies, a private company. Ormat adopted a systematic process for growth of its electricity production, from 8 MW to 110 MW, that enabled the gradual utilization of the geothermal reservoir's steam engines and minimized potential losses in the early, riskier years.
- The project was priced at USD 445 million to complete. After revision of the power purchase agreement (PPA) as well as the addition of a federal security plan to support the expenditures, the operator Kenya Power and Lighting Company, the venture was unable to attract the debt required for its growth until 2009.
- In the years 1998–1999, the private investor Ormat committed USD 40 million in equity investment, which grew to USD 150 million in 2006. As Ormat had to prolong its equity contribution further than anticipated, the company got the debt finance for just 11 years after the project began. The current project financing structure is mainly reliant on debt through Development Finance Institutions (DFIs) (accounting for 85% of overall investment costs). A funding committee headed by Germany's Deutsche Investitions- und Entwicklungsgesellschaft mbH (DEG) and KFW Development Bank refinanced Ormat's share in Phase I with a USD 105 million loan.
- The Overseas Private Investment Corporation (OPIC) of the United States granted a USD 310 million 19-year duration senior debt, payable in three tranches, to fund Phase II and Phase III construction and repay a portion of the debt and equity financing supplied previously.
- Olkaria III illustrates how a perfect blend of national support from the government, in the context of an exploration and grant, a security package going to guarantee electricity purchase, and an electricity purchase contract acknowledging core strategic risks, and international public financial services in the form of grants and Political Risk Insurance, can entice private capital into the geothermal industry in nations with limited access to capital.

References

1. Asian Development Bank, 2018, Philippines Energy Sector Assessment, Strategy and Road Map, Asian Development Bank, ISBN: 978-92-9261-362-4.

2. Chandrasekharam, D., 2015, Geothermal Energy Resources India: Country update, In *Proceedings of the World Geothermal Congress*, Melbourne, Australia, 19–25 April.
3. Fredleifsson, I.B., 2001, Geothermal Energy for the Benefit of People, *Renewable and Sustainable Energy Reviews*, Vol. 5 (2), pp. 299–312.
4. GSI, 1991, *Geothermal Atlas of India*, Geological Survey of India, Sp Pub, Hyderabad, Vol. 19, pp. 1–143.
5. Lund, J.W., Freeston, D.H., Boyd, T.L., 2005, Direct Application of Geothermal Energy, *Geothermics*, vol. 34, pp. 691–727.
6. Manikandan, S., 2015, Potential of Geothermal Resources in Power Generation in India, In *Proceedings of the World Geothermal Congress*, Melbourne, Australia, 19–25, April.
7. Mazzucalo, M., 2018, Financing Renewable Energy: Who Is Financing What and Why It Matters, *Technological Forecasting and Social Change*, Vol. 127, pp. 8–22.
8. Sircar, A. and Yadav, K. 2021, Geothermal Power Generation Potential of India, Unpublished report, pp. 1–28.
9. White, D.E., Muffler, L.J.P., Truesdell, A.H., 1971, Vapor-Dominated Hydrothermal Systems Compared with Hot-Water Systems, Economic Geology, Vol. 66, pp. 75–97.
10. World Bank, 2020, The Global Geothermal Development Plan: Mitigating Upstream Cost and Risk, The World Bank, IBRD-IDA.
11. Yadav, K. and Sircar, A., 2021, Geothermal Energy Provinces in India: A Renewable Heritage, *International Journal of Geoheritage and Parks*, Vol. 9 (I), pp. 93–107.
12. Yamamoto, T., Furuchata, T., Aranis, N. and Mori, K., 2001, Design and Testing of the Organic Rankine Cycle, *Energy*, Vol. 26 (3), pp. 239–251.

2

Enhanced Geothermal Systems

2.1 Introduction to EGS

The enhanced geothermal systems concept, which encompasses the older hot dry rock concept, was developed at the Los Alamos National Laboratory (LANL) in the United States. It was envisioned as a mass of hot rock formation located 5 kilometres beneath the earth's surface. The geothermal power stored in the heated rock was sufficient to generate electricity on the ground. Twenty years of studies at the Los Alamos Scientific Laboratory led in the discovery of a system that may be used to generate electricity at a low cost.

An EGS creates geothermal power without the use of mixed convection hydrothermal reserves (Abulkhaar et al., 2015). Until recently, geothermal technologies could only use resources with enough naturally produced heat, liquid, and rock porosity to generate heat. Furthermore, most of the geothermal power accessible by traditional means is trapped in dry, relatively impermeable formations. EGS techniques use a range of stimulation techniques, particularly "hydraulic stimulation," to increase and/or produce geothermal energy in these hot dry rock formations. If natural fractures, fissures, and openings do not permit adequate fluid velocity, the porosity of the rock must be increased by injecting high-pressure chilled water into the well. The infusion increases the hydrostatic flow in naturally cracked rock, causing shear processes that improve the permeability of the reservoir. A significant matrix porosity is not necessary as long as the infusion pressure is constant, and hydraulic fracking proppants are not required to maintain the cracks intact (Banjard et al., 2017). Hydro-shearing is the name given to this method to distinguish it from hydraulic tensile fracturing, which is employed in the oil and gas industry. This results in the formation of fresh cracks in the formation as well as the expansion of existing cracks. Enhancing permeability is a common practice (Chamorro et al., 2014). The chilled water passes through the rock's fissures and porous pores. It absorbs the heat of the rock and forces it out of a secondary borehole as extremely hot water. A heat engine or a binary power station system converts the heat from the water into power generation (Li et al.,

DOI: 10.1201/9781003204671-2

2014). In a closed system, all chilled water is pumped directly into the earth to warm again.

EGS systems can be used as power generators 24 hours a day, 7 days a week (Zemach et al., 2013). Like thermal energy, EGS can be used everywhere in the globe, dependent on the cost of drilling to a certain depth. Deeper granite with a 3–5-kilometre thickness of insulating deposits that restrict energy loss makes effective host formations. An EGS facility has a 20–30-year economic life cycle period.

In France, Australia, Japan, Germany, the United States, and Switzerland, EGS technologies are actively being implemented and deployed (Banjard, 2017). The largest global EGS venture is a 25-megawatt prototype plant now under construction in Australia's Cooper Valley. The Cooper Valley has the power to produce 5000–10,000 MW of electricity. In EGS, a certain level of induced seismicity is unavoidable and anticipated. This is due to the employment of hydro-shearing and hydraulic fracture processes, which entail injecting fluids at high pressure to improve or produce permeability. Hydro-shear activation techniques aim to improve the fluid system for the transmission of energy from bedrock to the fluid by expanding and extending the interconnectivity of the rock's pre-existing cracks. Seismicity episodes at California's Geysers thermal region have been found to be highly correlated with injection information (Abulkhaar et al., 2015). The instance of generated seismic activity in Basel is noteworthy; it prompted the city to halt the construction and undertake a seismic hazard assessment. The program was abandoned in December 2009 because of the assessment.

The hazards of "hydrofracturing produced seismicity are modest due to conventional earthquakes, and may be minimized by appropriate maintenance and coordination," as per the Australian government, and "will not be seen as an obstacle to ongoing exploitation of the Hot Rock geothermal power potential." The dangers of seismicity, on the other hand, vary from place to place and therefore should be evaluated before beginning large-scale fluid infusion.

2.1.1 Overview

Earth is a storehouse of thermal energy. The inner core is transferring heat to the outer core. Radioactive minerals in the subsurface generate heat. Researchers tried to gauge the amount of heat stored in the subsurface and found it to be enormous. An estimate of the total heat available in the earth's crust is about 540×10^7 EJ. Using 1% of this could sustain the energy needs of the earth for about 2500 years at a constant rate. However, the heat generated inside the earth is renewable in nature (Cheng et al., 2021). Direct application of geothermal energy has been available from prehistoric times in the form of hot springs and thermal pools. which have been used for cooking, bathing, and medical purposes. The use of geothermal energy first started about 10,000 years ago. However, the first geothermal heating system was installed in the centre of the French town of Chaudes-Aigues in the 14th century. At present, hot water and heat are extracted from the subsurface for direct and indirect uses. Electricity was first generated from geothermal steam in 1904 at a geyser in Italy. However, the first commercial generation of electricity took place in New Zealand in 1958. It used steam-operated turbines. Most of the geothermal energy extracted or pumped inside the earth has used a ground source heat pump (GSHP) (Gong et al., 2020). Hydrothermal resources are also used for electricity generation at a medium scale. However, this chapter is dedicated to EGS, which deals with extracting heat from hot dry rock. A chronology of efforts in this line is showcased in the subsequent section. The Harbanero plant (Australia) came into operation on 2 May 2013. It is the first commercial private plant to start generating electricity. Prior to this, there were only research plants, scattered across the globe, which were testing the efficacy of the new technologies.

2.1.2 Thermal Gradient

The geothermal gradient inside the ground, according to Bommer et al. (2006), varies around 25–30 °C per kilometre due to the heat of the earth. There are, nevertheless, several sites where the gradient is far higher. For instance, across a wide section of the western United States, the lowest temperature at a depth of 5 km is predicted to be 250 °C, implying a heat flux of at least 40 °C per kilometre of depth. A huge swath of western Canada is also ideal. Thermal energy can be classed as low, moderate, or high grade, depending on the anticipated heat flux of 30, 50, or 70 °C/km, respectively. This is supported by observations in southern Australia, where temperature of 273 and 283 °C were discovered, implying a thermal slope even higher than 50 °C per kilometre of depth (Yu et al., 2021). This is among the most attractive opportunities for an EGS facility, as per projections based on thermal variations, though it must be noted that such facilities are immensely complicated. Furthermore, some geological formations, such as significant thermal variations, are better suited to the utilization of geothermal power. For example, the areas of Alsace and Rhineland-Palatinate, where the Soultz-sous-Forêts and Landau EGS facilities are located, have a heat flux of up to 100 °C per kilometre of depth during the first 1000 metres. Substantial geothermal deposits that can be exploited for industrial energy generation are sought by industry. The very first task is to locate a reserve and collect the fluid within it so that the stored geothermal power can be transformed into power. The fundamental disadvantage is that such aquifers will only be used for a short time, that is, till the majority of the fluid within them has been removed. In this way, thermal deposits are comparable to oil resources in that they must first be discovered and then analysed to see if they hold enough fluid to make their production economical. After that, a facility must be constructed to use energy and transform it into electricity (Aliyn et al., 2021). It's been recommended that artificial water systems be developed to lessen reliance on naturally existing geothermal resources. Enhanced geothermal technologies, or EGS, is the name given to this approach. EGS is based on the idea of exploiting heat from "tight" rock that hasn't inherently fractured and has poor permeability. Activities are mostly centred on extracting heat from bedrock, utilizing freshwater as a working medium; to open the process, this combination is discharged onto hot rock regions (Lei et al., 2019). EGS weren't economically feasible initially, but technology developments in recent times have taken the concept closer to reality. Heat transfer from low-permeability geothermal systems is one of the ways that has received a boost. An important mission is to enhance the technologies needed to make EGS the huge energy source of the future (Liao et al., 2020).

2.2 Issues and Challenges for Subsurface Designing of EGS Systems

The high enthalpy accelerates downhole tubing corrosion and shortens the lifetime of other components exposed to drilling mud. To preserve the qualities required for drilling, the drilling mud should be maintained at a specified temperature. Fluid loss issues occur in all drilling operations, involving oil and gas, water systems, and geothermal boreholes. Bit and tool life is affected by three factors: Geological conditions, drilling parameters (including well path), and bottom hole installation. Table 2.1 represents the types of bits used for different types of geothermal drilling.

Annually, over 100 geothermal holes are drilled in the United States. Forecasts of deeper EGS reservoirs are hard to make because there are very few geothermal boreholes deeper than 2750 metres (9000 feet) in the United States. The budget of EGS drilling as a function of the distance can then be calculated using current geothermal production costs and oil and gas patterns. Ever since its development in the early 1970s, thermal exploitation has expanded in the United States, with a rush of activity in Northern California's The Geysers area, which would be a vapor-dominated steam source. Drilling issues at The Geysers were handled by inventing new oil

TABLE 2.1

Types of Bits for Geothermal Drilling

Bit Type	Issues and Challenges Faced during Drilling of Geothermal Wells	Geothermal Potential Zones
Roller cone bits	Seals and bearings	Commonly used for drilling in hard rocks
Polycrystalline diamond compact (PDC) bits	Hard formations	More effective with water jet
Water jet cutting technique	High pressure along with pumping seals	Limited
Percussion type	Good data and maintain gauge	Effective with fewer gauge problems
Flame jet type	Water quench, system difficulties, costs for deep wells	Hot dry rock reservoir (granite)

and gas techniques and adjusting existing techniques to the demanding downhole conditions of geothermal resources, such as cracked, tough, and brittle rocks, excess circulation loss, and higher temperatures.

2.3 Geothermal Drilling Technology for EGS

Geothermal drilling frequently employs the same cutting-edge equipment as the oil and gas industry, particularly in low-enthalpy sedimentary settings. Drilling is a critical tool for reservoir exploration and appraisal, as well as production and reinjection. Geothermal drilling has a number of distinguishing characteristics. Deep-seated, large-diameter boreholes and long-lasting well integrity are required for long-term geothermal development. Drilling for geothermal energy is more expensive. This is because geothermal reservoirs are more commonly found in igneous than sedimentary terrain. As a result, the rocks to be drilled are harder, more abrasive, and more fragmented than is found in the oil and gas sector.

2.3.1 Advanced Drilling Technologies

The following are examples of innovative technologies that could be applied to EGS: (i) evolutionary oil and gas well-drilling techniques that are now accessible and could be adapted to drill EGS reservoirs, and (ii) revolutionary technologies.

2.3.2 Evolutionary Oil and Gas Well-drilling Technologies

These comprise a variety of ways that can be used to bring down the cost of casing and cementing deeper EGS wells, hence considerably improving the profitability of deep EGS wells. The three most important principles, which deal with casing design, are broadly applied in the oil and gas sector and may be simply modified to EGS requirements. Multilaterals are utilized to decrease the price of accessibility to the reserve and have become standard practice in thermal and oil/gas production.

2.3.2.1 Expandable Tubulars

Along with the quantity of casing threads and the quantity of cement necessary, the expenses of enclosing a well with cement are significant. The use of extensible tubulars to line a well is a commonly available option. To assess the credibility of extensible tubular shells in wells where large temperature variation is envisaged, more research and testing are required. Efforts are being made to develop practical tools and specialist equipment to be used with expanding tubing, as well as to broaden the range of casing sizes available (Benzie et al., 2000; Dupal et al., 2001; Filippov et al., 1999). A bit precisely small enough to go just through to the thickest frame is used to drill the distance. The under-reamer is often utilized to enlarge the well's base and enable the casing to be cemented once it has run and expanded. The inside edges of neighbouring casings are flat as a consequence. This enables two main perspectives: (i) the resulting casing can be employed as the output string, and (ii) a liner can be run and placed in the well once the output interval has been finished. Technological advancements are necessary if this strategy is to be used for huge EGS reservoirs.

2.3.2.2 Under-reamers

Under-reamers are commonly used in oil and gas drilling into deposits because they give cement space for casing strings that would otherwise not be accessible. Monobore systems that use expanding tubulars need under-reamers. Bi-centre bits with polycrystalline diamond cutters are being used in oil and gas under-reaming. However, the effectiveness of polycrystalline diamond compact (PDC) blades in geothermal applications has yet to be determined. For EGS usage, more durable under-reamers are necessary.

2.3.2.3 Low-clearance Casing Design

Accepting decreased clearances can be used as an alternative to employing expanding tubulars. It may be acceptable to structure a well with shorter casing and less room between enclosing strings (Barker, 1997). An under-reamer may well be required to set up space between the casing and the well for cementing. While tighter specifications can present issues with cementing processes, this is normally avoided by using under-reamers prior to cement.

2.3.2.4 Drilling-with-Casing

This is an exciting technology with the ability to save money. This method may allow longer casing spacing, resulting in fewer strands – and hence lower costs (Gill et al., 1995). To enhance our understanding of cement methods that relate to the borehole approach, more research is needed. Advances in technological under-reamers, like flexible tubulars, are critical to the development of this technique.

2.3.2.5 Well Design Variations

Extending the duration of casing spacing can result in significant cost savings. As a result, the number of casing strands and, as a result, the width of the surface and first transitional casings will be reduced. The capacity to preserve wellbore stability of the drill span and construct a solid cement coating is important for the achievement of this strategy. There might be a few instances where this strategy is useful.

2.3.2.6 Multilateral Completions/Stimulating through Sidetracks and Laterals

Multilateral drilling and completions have advanced dramatically in the last 10 years. Nevertheless, if the most complex completing branch interconnections are being used, pressure-based activation of EGS aquifers may still be challenging.

2.3.3 Revolutionary Technologies

Drilling approaches may allow substantially better penetration rates and extended bit lifespan, lowering drilling costs. Spallation drilling, projectile drilling, laser drilling, and chemical drilling are examples of such procedures. High-enthalpy flames are used in spallation drilling to quickly heat the rock layer, forcing it to shatter or "spall." A technique like this might be used for melting non-spillable rock as well (Potter and Tester, 1998). Steel balls are fired at high speeds via pressured water to shatter and eliminate the rock layer in a projectile drill. The bullets are sorted from the drill mud and rock fragments and collected (Geddes and Curlett, 2006). Laser drilling employs the same method to extract rock, but it heats the rock layer using laser pulses. Chemical drilling has the potential of working in tandem with traditional drilling methods. It entails the use of powerful acids to dissolve the rock (Polizzotti et al., 2003). All these drill methods are still in the initial phases of development but are not yet widely viable. Nevertheless, effective development of any of these techniques could result in a significant shift in drilling procedures, much lower drilling costs, and, most importantly, the ability to drill deeper.

2.4 Power Production and Hydraulic Stimulation for EGS

Hydraulic activation is used to treat deep strata that already have fractures. These reservoirs should be under pressure, which will cause them to fracture and shear (Zhou et al., 2020). The following are the basic procedures for creating a reservoir:

- The infusion well must be drilled with casings that provides the average reservoir temperature.
- A thorough understanding of subsurface parameters such as load direction, joint measurement, in situ fluid description, rock geomechanical characteristics, and open zone data is required.
- Micro-seismic computed tomography must be placed in the best possible positions.
- Step mass flow injections until each injection step's pressure is stable.
- Ensure a constant flow rate until seismic activity moves to the next well.
- A shut-in test is required to determine the reservoir's size.
- The second well must be targeted up to the edge of the seismic activity substructure.
- The stimulation of the second well should be done in stages.
- To prove connection between infusion and production wells, interfering tests must be performed.
- The procedure might be replicated in the future to enable commercial electricity production.

2.4.1 Using Dense Fluid for Hydraulic Stimulation

Fluid stimulation is used to construct a subsurface heat exchanger in an EGS reservoir. Multiple paths are generated, and spontaneous cracks are stimulated, because of this process. The concern, however, is the emergence of micro-seismic events. Static pressure at the ground can be regulated by altering the volume of fluid pumped, the fluid injection times, and the viscosity of the material used for hydraulic stimulation, among other things. By optimizing the density of the working fluid being used by plant in the hydraulic stimulation process and adding solid solvents as needed, is the greatest way to control the constant capacity in the area that will also be fractured. It is feasible to (1) increase and improve subsurface circulation paths; (2) decrease the susceptibility of those pathways; and (3) enhance the mass flow and heat extractor ratios by using dense fluids (Olasolo et al., 2016). The fluid pressure at levels and regions where micro-seismic events occur could be monitored from the ground using sensors near the boreholes and computed using simulator programs. Due to the surface-related data, which is the most accurate data available, this software determines temperature and pressure through boreholes. The flow rate, pressure, density, and viscosity of the air are among the parameters used. At the Soultz-sous-Forêts facility, stimulation experiments were mostly undertaken in wells GPK2, GPK3, and GPK4. From 27 May 2003, a test run in well GPK3 found a collapse pressure difference of between 5 and 10 MPa/400 m. During the initial stages of the testing injection in wells GPK2 and GPK3, the infusion of saline (saturated NaCl) resulted in collapse pressures of 4.6 MPa/400 m in GPK2 and 4.4 MPa/400 m in GPK3 (Figure 2.1). At stability, it was anticipated that there would be a substantial rise in concentration with depth in both boreholes (Fritz et al., 2010). Using the results of experiments conducted in September 2004, this hypothesis was confirmed in well GPK4 (Heidinger et al., 2010). From the day of the injection, the differential pressure in the open part of the infusion well dropped in both boreholes, leading to a decline in the volume of the fluid surface as it moved deeper into the well. The rise in density induced by the drop in operating temperatures is outweighed by such an impact. Due to the high density of the fluid injected, the heat differential between the top and the base of the well rises when the injection carrying a level of NaCl hits the bottom of the well (Zimmermann et al., 2010).

2.4.2 Using Carbon Dioxide (CO_2) as a Working Fluid

The idea of using CO_2 instead of water as a working fluid at high pressures introduces a new concept in the operation of EGS plants. The thermal/physical properties of CO_2 make it an attractive option as a heat transfer medium. If brine is used as a working fluid, water reacts with minerals within the reservoir, which causes challenges (Frash et al., 2015). CO_2 causes no appreciable dissolution or precipitation problems.

Table 2.2 shows an outline comparison between CO_2 and water as working fluids in EGS plants. In an EGS plant the viscosity of CO_2 increases

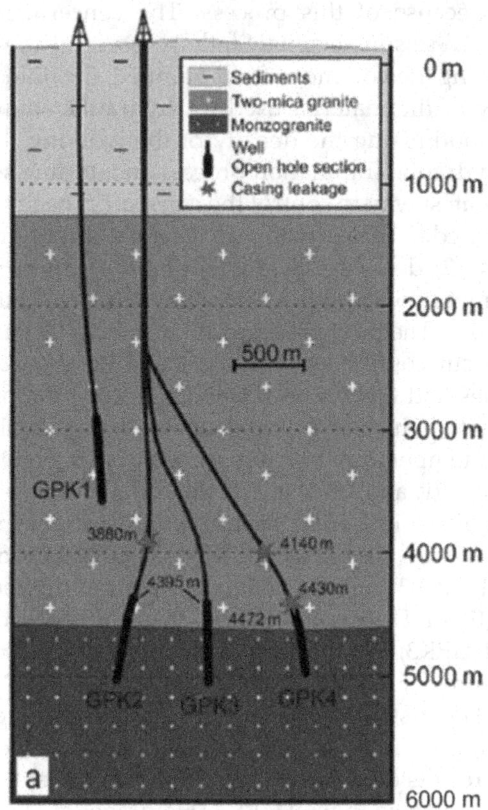

FIGURE 2.1
Well trajectories of GPK-1, GPK-2, GPK-3, and GPK-4 shown through the subsurface.

TABLE 2.2
Comparison of CO_2 and Water as Working Fluids in an EGS Type Geothermal Plant

Fluid Properties	Carbon Dioxide (CO_2)	Water
Chemical	Ionic dissolution product (No) An ionic dissolution product (serious) Mineral dissolution/precipitation problems	Problems of mineral dissolution/precipitation
Fluid circulation	Highly compressible and expandable	Low compressibility (moderate)
Ease of flow in the geothermal reservoir	Low viscosity and density	High viscosity and density
Heat transmission	Low specific heat level	High specific heat level
Fluid losses	Could result in beneficial geological storage of CO_2	A hindrance to the development of geothermal reservoirs (costly)

much less at low temperatures, so most of the pressure is available for use in the production well.

2.4.3 Software Packages Used for EGS Investigation

EURONAUT is a European modelling software suite for EGS. This software suite now makes it possible to perform impact assessments on EGS with confidence. It was developed based on research undertaken at the EGS facility in Soultz-sous-Forêts. Economic assessment using interrupted cash flows is the focus of the package. The software contains several components that can be linked together via an application. On the basis of knowledge gained in Soultz-sous-Forêts, multi-well installations, such as the present EGS facility in Soultz-sous-Forêts, are treated as existing wells. Every well's data is analysed separately or in unison (Pollack et al., 2019).

2.5 Influence of Micro-seismic Events in EGS

An algorithm based on short-term wavelength to long-term wavelength gives you amplitude ratios. Then, for P and S waves, a beam forming algorithm should be used.

An event is assigned once a given number of detections are found for a predefined grid in the subsurface fulfilling minimum travel time residuals (Calò et al., 2011). The second step determines the P-wave onset time by application of the Akaike information criterion, followed by a polarization analysis. The third step uses the onset times and polarization angles together with user-defined weights on the different data to invert for the location using a directed arid search and pre-calculated travel timetables. After all data have been processed, reprocessed, and often manually quality controlled, the final task is interpretation of the results. To simplify the interpretation of the hydraulics stimulation, micro-seismic event locations should be visualized together with all other relevant events (Evans et al., 2012). In addition, the origin time of the events, their magnitudes, and the locations' uncertainty estimates are crucial in the visualization.

2.5.1 Micro-seismic Events and Stimulation of Enhanced Geothermal System

Prior to hydraulic fracturing, a micro-seismic network is installed in enhanced geothermal sites. This monitors background seismicity in the area. Borehole geophones are planted in the shallow and deep levels of the drilled hole. In most geothermal modelling, the velocity modelling information is stored in the form of data. Prior to drilling in modern geothermal sites, seismic data is either acquired or reprocessed to augment stimulation. The subsurface fault skeleton is mapped from the seismic data. An understanding of subsurface features is necessary before stimulation. During stimulation for 4–5 days, 7000–10,000 events are normally detected and located in real time. The first step is a multichannel detection.

References

1 Abulkhaar, H., Crooke, D., Hawd, M., 2015, Seismic Mapping and Geochemical Analyses of Faults within Deep Hot Granites, a Workflow for Enhanced Geothermal System Projects, *Geothermics*, Vol. 53, pp. 46–56.

2 Aliyn, M.D., Archer, R.A., 2021, Numerical Simulation of Multi fracture HDR Geothermal Reservoirs, *Renewable Energy*, Vol. 164, pp. 541–555.

3 Banjard, C., 2017, Hydrothermal Characterization of Wells GRT-1 And GRT-2 in Rittershoffen, France: Implications on the Understanding of Natural Flow Systems in Rhine Graben, *Geothermics*, Vol. 65, pp. 255–268.

4 Barker, J.W., 1997, Wellbore Design with Reduced Clearance between Casing Strings, In *Proc. SPE/IADC Drilling Conference*, SPE/IADC 37615, Amsterdam, Netherlands, March 1997. https://doi.org/10.2118/37615-MS.

5 Benzie, S., Burge, P., Dobson, A., 2000, Towards a Mono-diameter Well – Advances in Expanding Tubular Technology, In *Proc. SPE European Petroleum Conference 2000*, SPE 65164, Paris, France, SPE paper #65184.

6 Bommer, J.J., Oates, S.J., Cepeda, M., Lindholm, C., Bird, J., Torres, R, Marroquin G., Rivas,J., 2006, Control of Hazard Due to Seismicity Induced by a Hot Fractured Rock Geothermal Project, *Engineering Geology*, Vol. 83, pp. 287–306.

7 Calò, M., Dorbath, C., Cornet, F., Cuenot, N., 2011, Large-scale Aseismic Motion Identified through 4D P-wave Tomography, *Geophysical Journal International*, Vol. 186, pp. 1295–1314.

8 Chamorro, C.R., Garcia-Cuesta, J.L., Mondejar, M.E., Perez-Madrazo, A., 2014, EGSs in Europe; an Estimation and Comparison of the Technical and Sustainable Potentials, *Energy*, Vol. 65, pp. 250–263.

9 Cheng, C., Hermann, J., Rybacki, E., Milsch, H., 2021, Long-term Evolution of Fracture Permeability in Slate as Potential Target Reservoirs for Enhanced Geothermal Systems (EGS), EGU21-12187, EGU General Assembly 2021, updated on 4 March, 19–30 April, 2021. https://doi.org/10.5194/egusphere-egu21-12187, EGU General Assembly 2021.

10 Dupal, K.K., Campo, D.B., Lofton, J.E., Weisinger, D., Cook, R.L., Bullock, M.D., Grant, T.P., York, P.L., 2001, Solid Expandable Tubular Technology – A Year of Case Histories in the Drilling Environment, *Proc. SPE/IADC Drilling Conference 2001*, SPE/IADC 67770, Amsterdam, Netherlands. doi:10.2118/67770-MS.

11 Evans, K.F., Zappone, A., Kraft, T., Delchmann M., Mola, F., 2012, A Survey of the Induced Seismic Responses to Fluid Injection in Geothermal and CO_2 Reservoirs in Europe, *Geothermics*, Vol. 41, pp. 30–54.

12 Filippov, A., Mack, R., Cook, L., York, P., Ring, L., McCoy, T., 1999, Expandable Tubular Solutions, SPE Annual Conference and Exhibition, SPE 56500, Houston, Texas, October 1999, Paper Number: SPE-56500-MS. https://doi.org/10.2118/56500-MS.

13 Frash, L.P., Gutierrez, M., Hampton, J., Hood, J., 2015, Laboratory Simulation of Binary and Triple Well EGS in Large Granite Blocks Using AE Events for Drilling Guidance, *Geothermics*, Vol. 55, pp. 1–15.

14 Fritz, B., Jacquot, E., Jacquemont, B. Baldeyron-Bailly, A., Rosener, M., Vidal, O., 2010, Geochemical Modelling of Fluid–Rock Interactions in the Context of the Soultz-sous-Forêts Geothermal System, *Comptes Rendus Geoscience*, Vol. 1342 (7–8), pp. 653–657.

15 Geddes, C.J., Curlett, H.B., 2006, Leveraging a New Energy Source to Enhance Oil and Oil Sands Production, *GRC Bulletin*, January/February, pp. 32–36.

16 Gill, D.S., Lohbeck, W.C.M., Stewart, R.B., van Viet, J.P.M., 1995, Method of Creating a Casing in a Borehole, Patent No. PCT/EP 96/0000265, Filing date: 16 January 1995.

17 Gong, F., Guo, T., Sun, W., Li, Z., Yang, B., Chen, Y., Qu, Z., 2020, Evaluation of Geothermal Energy Extraction in Enhanced Geothermal System (EGS) with Multiple Fracturing Horizontal Wells (MFHW), *Renewable Energy*, Vol. 151, pp. 1339–1351.

18 Heidinger, P., 2010, Integral Modeling and Financial Impact of the Geothermal Situation and Power Plant at Soultz-sous-Forêts, *Comptes Rendus Geoscience*, Vol. 342, pp. 626–635.

19 Lei, Z., Zhang, Y., Yu, Z., 2019, Exploratory Research into Enhanced Geothermal System Power Generation Project, The Qiabugia Geothermal Field, Northwest China, *Renewable Energy*, Vol. 139, pp. 52–70.

20 Li, M.Y., Lior, N., 2014, Comparative Analysis of Power Plant Options for Enhanced Geothermal System, *Energies*, Vol. 7 (12), pp. 8427–8445.

21 Liao, Y, Sun, X, Sun, B., 2020, Geothermal Exploitation and Electricity Generation from Multibranch U-shaped Well Enhanced Geothermal System, *Renewable Energy*, Vol. 151, pp. 782–795.

22 Olasolo, P., Juarez, M.C., Morales, M.P., D'Amico, S., Liarte, L.A., 2016, Enhanced Geothermal Systems (EGS): A Review, *Renewable and Sustainable Energy Review*, Vol. 56, pp. 133–144.

23 Polizzotti, R.S., Hirsch, L.L., Herhold, A.B., Ertas, M.D., 2003, Hydrothermal Drilling Method and System, United States Patent No. 6.742,603, 3 July.

24 Pollack, A., Mukherji, T., 2019, Accounting for Subsurface Uncertainty in Enhanced Geothermal Systems to Make More Robust Techno-economic Decisions, *Applied Energy*, Vol. 254, 113666.

25 Potter, R.M., Tester, J.W., 1998, Continuous Drilling of Vertical Boreholes by Thermal Processes: Including Rock Spallation and Fusion, United States Patent No. 5,771,984, 30 June.

26 Yu, G., Lin, C., Zhang, L., Fang, L., 2021, Parameter Sensitivity and Economic Analysis of an Interchanged Fracture Enhanced Geothermal System, *Advances in Geo-Energy Research*, Vol. 5 (2), pp. 166–180.

27 Zemach, E., Drakos, P., Spielman, P., Akerley, J., 2013, Desert Peak East Enhanced Geothermal Systems (EGS) Project. United States: N. 2013. Web. doi:10.2172/1373310.

28 Zhou, Z., Jun, Y., Zeng, Y., 2020, Investigation on fracture creation in hot dry rock geothermal formations of China during hydraulic fracturing, *Renewable Energy*, Vol. 153, pp. 301–313.

29 Zimmermann, G., Reincke, A., 2010, Hydrantic Stimulation of a Deep Sandstone Reservoir to Develop an Enhanced Geothermal System: Laboratory and Field Experiments, *Geothermics*, Vol. 39 (1), pp. 70–77.

3

Direct Applications: Application of Geothermal Water for Sericulture, Aquaculture, and Honey Processing

3.1 Direct Utilization of Geothermal Energy around the World

Around the world, geothermal energy is primarily used for societal applications such as heating aquaculture ponds and raceways, snow melting, space cooling, and balneology, among others. The utilization of geothermal energy in fish farming has increased 36.5% in terms of installed capacity and 13.5% in terms of annual energy use since 2015 (Darma et al., 2020), with a maximum capacity of 950 MWt and an annual power usage of 13,573 TJ. Israel, Iceland, Italy, China, and the United States accounted for 92% of annual revenue in 2015 (Ben Mohammed, 2015). Major energy consumption among these 21 countries is primarily for power generation. These services, including greenhouses, are labour intensive and require well-trained workers. Development is slow because it is hard to justify on a financial level. Salmon, trout, bass, and tilapia are among the most common species; however, shrimp, prawns, lobsters, tropical fish, and alligators are also farmed using geothermal water (Tian et al., 2020). Colorado and Idaho are the two common alligator breeding facilities in the United States using geothermal water, where the alligators are grown for meat and hides. They are raised as a tourist attraction. When using geothermal energy in exposed ponds, 0.242 TJ/yr/tonne of fish is needed, according to research conducted in the United States on bass and tilapia. As a result, the recorded annual energy consumption of 13,573 TJ corresponds to a projected annual production of 56,087 tonnes, an improvement of 13.5% over 2015.

It's worth remembering that the energy requirements are about half as high if the fish are raised in protected ponds, such as in a greenhouse. However, there are only a few known protected ponds in use (Fričovský et al., 2020). These are the data that are the most challenging to obtain and measure. Every other country (53 of 88) has spas and resorts with geothermal-heated swimming pools, like balneology (water-based

DOI: 10.1201/9781003204671-3

medicine), but many never regulate the water flow except at night. Many nations do not keep track of how much time people spend in pools. As a result, the real consumption and capability estimates could be off by up to 20% (Morozov et al., 2020). Capacity and energy use calculations of 0.35 MWt and 7.0 TJ/yr were used where no flow or temperature drop was observed. Focusing on data from different countries, 5 l/s with a 10°C temperature change was used in the generation potential of 0.21 MWt, and 3 l/s with a 10°C temperature increase was taken for the use of 4.0 TJ/yr (Wakana, 2020).

In comparison to the 53 nations that registered balneology use, there are reported facilities in Zambia, Singapore, Denmark, Mozambique, Nicaragua, and France for which no data is available. Balneology uses the most power, accounting for 79.5% of all annual energy use in China, Japan, Turkey, Brazil, and Mexico.

In Iceland, Japan, Argentina, and other countries, snow melting solutions for roads and sidewalks are accessible (Yasukawa et al., 2020). This includes the United States and Slovenia, with Poland and Norway following suit to a lesser degree. Approximately 2.5 million square metres are heated all around world, with Iceland accounting for the bulk at around 74% (Kępińska, 2020). A project in Argentina's Andes utilizes geothermal energy to remove road snow and keep resorts open throughout the year (Chiodi et al., 2020). The majority of road snow removal in the United States takes place on the Oregon Institute of Technology campus and in Klamath Falls, where it can be used as part of a district heating system that uses lower-temperature return water in a heat exchanger with a glycol–water mixture to melt snow from roads and bridge decks (Ragnarsson et al., 2020). The required power ranges from 130 to 180 W/m^2 for the United States and Iceland, respectively. Snow melting capacity is 415 MWt, with an annual energy consumption of 2389 TJ. With a maximum of 19.9 MWt and 200.1 TJ/yr. Bulgaria is the largest user of air conditioning, followed by Slovenia, India, Brazil, Australia, and Algeria (Chandrasekharam and Chandrasekhar, 2020). Land source heat pumps are not included because they only transfer heat to the ground and thus do not use geothermal energy. Both snow melting and air conditioning are estimated to use 434.9 MWt and 2589.1 TJ/yr.

3.2 Application of Geothermal Water for Sericulture

Silk has been an intellectual endeavour for decades, and enhanced analysis tools and molecular biology techniques are revealing new information about these polymers. Silk is a protein-rich high-molecular-mass polymer that is often associated with insects, silkworms, and orb weaving spiders (Raje and Rekha, 1998). The tiny caterpillar of the silk worm *Bombyx mori*, is the key to silk development. It only eats the leaves of the mulberry tree (Chappell et al., 2013). Only one other insect species, *Antheraea mylitta*, develops silk fibre. The silk filament of this wild creature is approximately three times stronger than that of cultivated silkworms (Maji et al., 2008). The coarser fibre is known as tussah. The life cycle of *Bombyx mori* begins with the adult insect laying eggs. The larvae hatch from the eggs and develop by eating mulberry leaves. In nature, the chrysalis emerges as a moth after breaking through the cocoon (Bandhopadhyay and Kumar, 2013).

After breeding, the female insect lays 300 to 400 eggs. After emerging from the cocoon, the insects perish a few days later, and the life cycle starts again. Sericulture is the breeding of silkworms for the manufacture of silk. Over the ages, sericulture has been developed and improved into a precise science (Miyashita et al., 1999). Sericulture refers to the procedure of raising healthy worms all the way to the chrysalis phase, where the worm is wrapped in its silky nest. The silk thread is preserved by killing the worm within the cocoon before it can move away. Only the best health insects are selected for reproduction, and they are given the opportunity to grow, mate, and lay other eggs. In general, one cocoon yields 1000 to 2000 feet of silk filament, which is composed primarily of two components (Engel and Hoppe, 1988). The fibre, known as fibroin, accounts for 75–90% of silk. Sericin, the glue naturally produced by the moth to glue the fibre into a chrysalis, contributes 10–25% of the total. Other ingredients include fats, salts, and wax. Approximately 3000 cocoons are used to produce 1 yard of silk fabric (Hopwood et al., 2005).

Sericin is a common type of silk material that comprises 18 amino acids, particularly essential amino acids and 32% serine. The total percentage of hydroxy amino acids in sericin is 45.8%. In total, 42.3% of the amino acid residues are polar, while 12.2% are non-polar (Carew and Mitchell, 2008). Sericin accounts for about 20–30% of the overall cocoon weight, primarily encircling fibroin. When sericin is present, the polymers are hard and durable, but when it is removed, they become soft and lustrous. Sericin is composed of amorphous irregular coils and, to a lesser extent, beta-sheet ordered structure. As a result of repetitive moisture absorption and mechanical spreading, the randomly coiled structure quickly transforms into beta-sheet structure. The solubility of gum in hot water (50–60 °C) is solely responsible for its removal from crude silk (Yamada et al., 2001).

There are a variety of methods for purifying silk:

1. Gum extraction with a dilute sodium carbonate solution.
2. Raw silk is extracted using hot water, then evaporated to produce powder.

3. Boiling the raw silk in water and re-boiling till the extract no longer gives a gallic acid precipitate.
4. Three cycles of 1-hour silk extractions, or simply heating in water at 100 °C, pasteurization at 118 °C, or autoclaving for 3 hours at 2.5–3 times atmospheric pressure.
5. Sericin, with an estimated molecular mass of 50,000, can be obtained from cocoons using an aqueous urea mixture at 100 °C.
6. To prevent decomposition, use water at 50–60 °C for 25 days.
7. Boiling formulations of pH 11 comprising 5–6% bentonite will completely degum silk fibres.

The bulk of sericin can be removed by autoclaving for 1.5 hours at 600–700 mmHg (14 lb) pressure, although increasing the treatment time to 2 hours induces no bigger loss in fibre weight. It was also found that extracting sericin at low pressure (25 cm Hg, 5 lb) produces satisfactory results. Sericin is extracted from *Bombyx mori* cocoons using a water bath and pasteurization at different temperatures; the best production is obtained by pasteurization at 105 °C for 30 minutes, which has excellent gelling properties and yield. The yield increases as the temperature increases, but the gelling capacity is lost (Yoshioka et al., 2001).

Enzymes are used to extract silk sericin. Removal can be done at 60 °C for 90 minutes at pH 10 with enzyme alkylase or 2–2.5 g/L alkaline protease. Sericin is extracted using trypsin hydrolysis at various concentrations, temperatures, and treatment times. At 37 °C and 20 °C, hydrolysis of 1% trypsin solution is almost complete in 10 and 32, respectively. After 4 hours of treatment with 1% and 8% trypsin solution, the amount of sericin provided is 26.4% and 28.7%, respectively (Good et al., 2009).

3.2.1 Steps of Processing through Geothermal Water

For breeding, only the healthiest moths are included. Caterpillars spin silk around themselves, finally fully encasing themselves. The cocoons are soaked in geothermal water to free the sericin stuck in the fibres after being dissolved to destroy the chrysalis. The silk fibre is woven through an eyelet and reeled to a wheel at the other end. The silk thread is twisted after being washed with oil or soap and dried. The texture of the fabric is determined by the type of twisting used. The stages of silk processing through geothermal water are depicted in Figure 3.1.

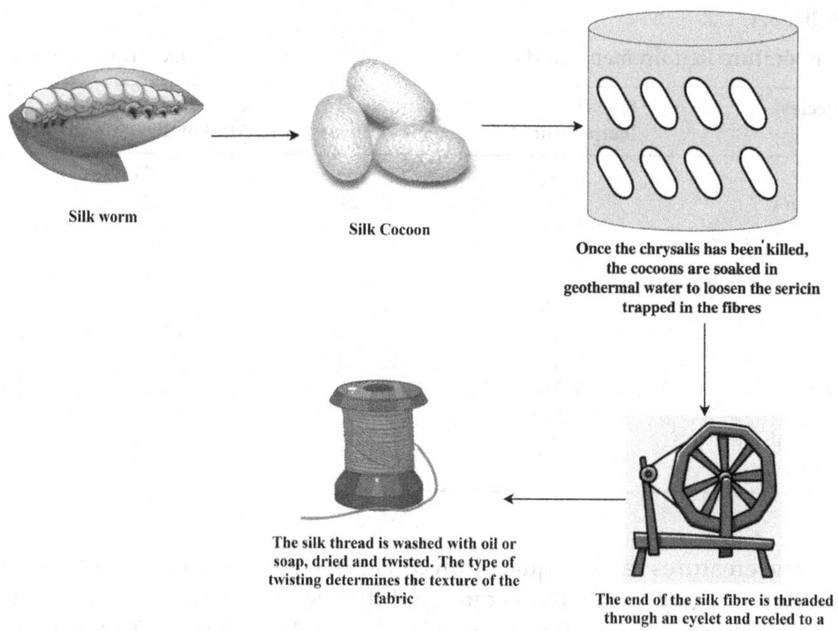
Silk worm
Silk Cocoon
Once the chrysalis has been killed, the cocoons are soaked in geothermal water to loosen the sericin trapped in the fibres
The end of the silk fibre is threaded through an eyelet and reeled to a wheel
The silk thread is washed with oil or soap, dried and twisted. The type of twisting determines the texture of the fabric

FIGURE 3.1
Stages involved in silk processing through geothermal water.

3.3 Application of Geothermal Water for Aquaculture

Aquaculture is the growing of food (fish and molluscs) and nutrients in regulated water sources (e.g. algae). Many species will grow faster and bigger in warmer temperatures. The temperature of fish production plants may be controlled using geothermal waters, allowing the production of bigger, faster-growing fish as well as fish production during the winter months when it would otherwise be impossible. In most cases, geothermally heated water may be directly used in ponds. Heat exchange equipment is no longer needed.

There are no commercially viable energy options for aquaculture that uses geothermal resources. Only a geothermal energy-heated plant will provide low-cost, safe hot water that will enable the fish to grow rapidly and survive the winter if a farm needs to produce a specific species in a specific climate (Taylor, 2005). Some people heat their ponds with natural gas or propane, but this is not a cost-effective option, since it necessitates the procurement of a costly burner and furnace, as well as energy.

TABLE 3.1

Temperature Requirements and Growth Periods for Selected Aquaculture Species

Species	Tolerable Temperature (°C)	Optimum Growth Temperature (°C)	Growth Period of Market Size (months)
Oysters	0–36	24–26	24
Lobsters	0–31	22–24	24
Shrimps	4–40	25–31	6–8
Salmon	4–25	15	6–12
Freshwater Prawns	24–32	28–31	6–12
Catfish	17–35	28–31	6
Eels	0–36	23–30	12–24
Tilapia	8–41	22–30	–
Carp	4–38	20–32	–
Trout	0–32	17	6–8
Yellow Perch	0–30	22–28	10
Striped Bass	0–30	16–19	6–8

Some creatures can be aquacultured without geothermal energy (due to the fact that geothermal use is rare), but the regions are limited due to climate, and the growth rates are slower (Lund, 2010). Table 3.1 shows the temperature requirements and development periods for different aquaculture species.

Raceway wall heat transfer, raceway water heat transfer, radiation temperature distribution from the water surface to the surroundings, heat resistance, and gross thermal dissipation per month are only a few of the variables to consider (Hamza's Reef, 2016). The techniques for designing the aquaculture facility are as follows.

3.3.1 Stage 1: Estimation of Heat Transfer between the Water Channel and the Walls

The sum of heat transfer through the lateral concrete walls in a confined water channel should be calculated using both displacement heat transfer process and conduction. The linear conduction principle governs the transfer of energy (Murray, 2014; Saeedi et al., 2019).

For each month's assessment of conductivity-based heat transfer resistance of the water channel's concrete walls, Eq. (3.1) was used.

$$R_1 = \frac{l_{cement}(t_w)}{K_{cement} \times A_{wall}} \tag{3.1}$$

The resistance to conduction through concrete barrier (in K/W) is R_1, and the concrete block width (m) is l_{cement} (t_w). The void separating two walls is referred to as $A_{wall}(m^2)$. For a turbulent flow, the hydraulic diameter can be used to measure the Reynolds number. A dimension to remember for open water channels or non-circular parts is length or dynamical diameter. This volume is used when the fluid is circulating in a non-circular section. The hydraulic diameter, Reynolds number, and Prandtl number were calculated using relationships Eq. (3.2), (3.3), and (3.4).

$$D_h = \frac{4 \times a \times b}{(2 \times a) \times b} \quad (3.2)$$

Where D_h signifies the water channel's hydraulic diameter (m), a denotes the raceway's wall height (m), and b denotes the water channel's width (m).

$$Re_w = \frac{D_h \times \rho_w \times V_w}{\mu_w} \quad (3.3)$$

Where Re_w is the monthly Reynolds number of water, w is the monthly density of water (kg/m^3), $\rho_w = \mu_w$ is the monthly viscosity (Pa.s), and V_w is the monthly velocity of water movement (m/s).

$$P_{rw} = \frac{D_h \times K_w \times C_{pw}}{\mu_w} \quad (3.4)$$

While the fluid is flowing, heat transfer is necessary, and once the fluid is not moving, it is a form of conduction. In a turbulent flow with a flat surface, the Nusselt value is used to estimate the movement boundary between the stream and the wall.

Eq. (3.5) uses the Nusselt value to quantify heat flow from a horizontal water surface in an aquatic ecosystem to the ambient air. The estimated number derived for measuring the coefficient of heat flow can be used as a kind of free displacement (Antoine-Frostburg State University, 2016; Murray, 2014).

$$Nu_w = 0.0296 \times (Re_w)^{0.8} \times (Pr_w)^{0.333} \quad (3.5)$$

The Nusselt number for each month is denoted by Nu_w. Eq. (3.6) calculates the heat flow coefficient by using the Nusselt value and the thermo-physical characteristics of water that are used to quantify heat resistance caused by heat transfer. To look at it differently, heat transfer due to free displacement from the surface of the water to the air is calculated using Eq. (3.7), in other words, the resistances against transfer relation (Bergman et al., 2011; Murray, 2014).

$$hw = \left(NuNu_w \times Kw\right) / D_h \tag{3.6}$$

Where hw represents the monthly heat flow coefficient of water (W/m².K).

$$R_2 = \frac{1}{h_w \times A_{wall}} \tag{3.7}$$

Where, for every month (K/W), R_2 signifies heat resistance due to convection.

Eqs. (3.8) and (3.9) were used to obtain the total heat transfer resistances and heat transfer of the entire device from the wall of a raceway for each month, which are obtained by aggregating conduction and displacement heat transfers and resistances using the raceway's data and metrics.

$$R_{wall} = R_1 + R_2 \tag{3.8}$$

Where R_{wall} indicates the total wall's resistance to heat transfer (K/w).

$$q_{wall} = \frac{\left(T_{w1} \times T_{inf}\right)}{R_{wall}} \tag{3.9}$$

For each month (W), q_{wall} signifies heat transfer through the raceway's side walls, and T_{w1} reflects the upper bound water temperature (K).

$$T_{inf} = 0.0552 \times T_a^{1.5} \tag{3.10}$$

In Eq. (3.10), T_{inf} (Cameron Morton, 2013 and Garg et al., 2000) represents the mean sky temperature for each month (k), and T_a signifies the month's normal air temperature (k) (Cameron Morton, 2013; Garg et al., 2000).

3.3.2 Stage 2: Heat Transfer Estimation of Water in Raceway

For each month, the heat flow and resistance to heat flow of free movement of water into the air are calculated using Eqs. (3.11) and (3.12), which is expressed as Eq. (3.19) (MHTL, 2016).

$$B = \frac{1}{T_{inf}} \tag{3.11}$$

Every month's Biot number is represented by B.

$$L_c = \frac{a \times b}{2 \times (a \times b)} \tag{3.12}$$

Where L_c signifies characteristic length (m).

$$Gr_a = \frac{g \times B \times (T_{w1} - T_{inf}) \times (L_c)^3}{(V_a)^2} \tag{3.13}$$

Where Gr_a stands for Grashof amount of air per month, g signifies gravity's acceleration, which is 9.80665 m/s², and V_a represents air's kinematic viscosity (m²/s) (MHTL, 2016).

$$Pr_a = \frac{C_{pa} \times \mu_a}{K_a} \tag{3.14}$$

Where Pr_a denotes the Prandtl value of air for each month, C_{pa} denotes the specific heat of air (j/kgK), μ_a denotes the viscosity of air (Pa.s), and K_a denotes the heat capacity of air (W/mK).

$$Ra_a = Gr_a Pr_a \tag{3.15}$$

Where Raleigh is represented by Ra_a.

$$Nu_a = 0.15 \times (Ra_a)^{0.333} \tag{3.16}$$

Where Nu_a signifies the Nusselt value of air for each month.

$$h_a = \frac{Nu_a \times K_a}{L_c} \tag{3.17}$$

Where h_a represents air heat transfer coefficient every month (W/m²K).

$$Rf_{conv} = \frac{1}{h_a \times (a \times b)} \tag{3.18}$$

Where Rf_{conv} represents heat resistance through water surface due to free convection (K/W).

$$q_{fconv} = \frac{\left(T_{w1} - T_{inf}\right)}{R_{fconv}} \tag{3.19}$$

Where heat flow through the water surface due to free convection is denoted by q_{fconv} (W).

3.3.3 Stage 3: The Heat Flow from the Water Surface to the Atmosphere Is Calculated Using Radiation

The amount of radiation heat flow from the water surface in water channel aquaculture is calculated by the surrounding area. Eq. (3.20) shows how to quantify total radiation heat transfer:

$$q_{rad} = \varepsilon_w \times (a \times b) \times \delta_b \times \left(T_{w1}^4 - T_{inf}^4\right) \tag{3.20}$$

As seen, total radiation heat transfer can be measured. Where q_{rad} is the heat transfer coefficient from the water surface caused by radiation (W), w is the constant coefficient of 0.98, and δ_b is the Boltzmann constant of 5.67 × 10–8 W/m^{2}K^4.

3.3.4 Stage 4: Heat Resistance and Total Heat Dissipation Estimation Each Month with Only Water Condition

The three mechanisms that make up the basic framework are conduction heat flow, relocation, and radiation. Heat flow, permanent and ordered thermal dissipation, and accumulation of these three key mechanisms are established in aquaculture. Without use of biomass, the q_{base}, or central heat load, can keep the water at a constant temperature. Eqs. (3.21) and (3.22) were used to determine the thermal transfer resistance of the entire device as well as q_{base} and R_{base} (Bergman et al., 2011; Murray, 2014).

$$q_{base} = q_{wall} + q_{fconv} + q_{rad} \tag{3.21}$$

Where, for each month, q_{base} denotes complete heat transfer out of water-only baseline method (W).

$$R_{base} = \frac{\left(T_{w1} - T_{inf}\right)}{q_{base}} \tag{3.22}$$

Where R_{base} represents the month's heat resistance of a water-only baseline device (K/W).

3.3.5 Stage 5: Calculation of the Time Taken to Cool or Heat the Device without the Use of Biomass

Eq. (3.23) and the details in Table 3.1 relating to the lower limit water temperature have been used to calculate the approximate heat resistance (Eq. (3.34)) and visualize the impact of biomasses on the time taken to cool/heat the system by adding biomass to the aquatic setting. From the temperature of each month, the ambient temperature for that month, and heat resistance (R_{base}) per hour, this equation calculated the time it took to heat or cool water without biomass. The span of time it takes to cool or heat base water (without biomass) at room temperature per month is represented by t_w in Eq. (3.23).

$$t_w = \rho_w \times V_{rw} \times C_{pw} \times R_{base} \times \ln \frac{\left(T_{w1} - T_{inf}\right)}{\left(T_{w2} - T_{inf}\right)} \tag{3.23}$$

Where T_{w2} signifies lower bound water temperature (K) and t_w denotes time (h) to cool/warm water-only system.

3.3.6 Stage 6: Estimation of Heat Flow from the Biomass by Conduction and Displacement

When a constant amount of biomass is used in aquaculture, a portion of the volume of water is substituted with solid fish biomass, which has different thermo-physical characteristics than water.

When biomass is connected to the network, it produces a surplus heat flow mechanism between the aquatic organisms and the water used in the basic system. A part of the heat transfer diagram from aquatic ecosystem to biomass is shown in Figure 3.2.

Specific fish heat flow to the system is depicted as a single heat resistance in this simple diagram (R_{fish}) (Murray, 2014). Eqs. (3.24) to (3.32) are used to calculate the thermal resistance of biomass.

$$SA_t = P_t \times b \tag{3.24}$$

Where SA_t signifies the aquatic life dedicated section (m^2) and P_t denotes the fish's length (m), which equals (Brautigam, 2015)

$$P_t = 2\pi \times r_t \tag{3.25}$$

Where the radius of the trout is denoted by r_t (m).

$$DH_t = \frac{\left(4 \times A_t\right)}{P_t} \tag{3.26}$$

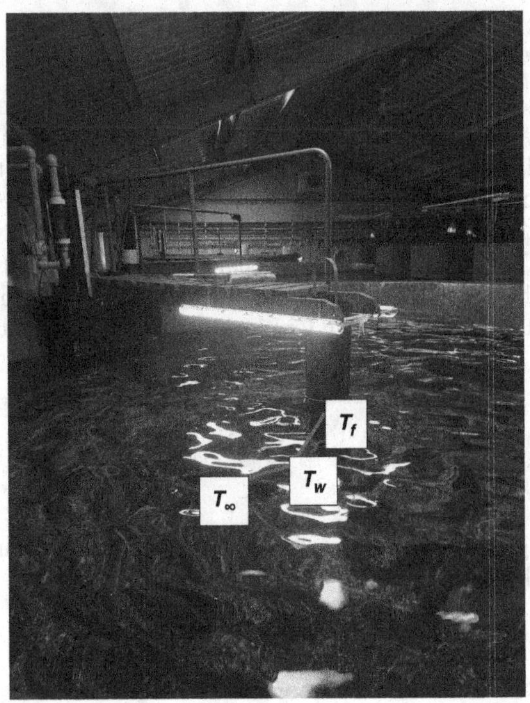

FIGURE 3.2
Simplified heat transfer diagram – biomass system.

Where DH_t signifies the aquatic species' hydraulic diameter (m) and A_t signifies the trout's area (m²).

$$A_t = (r_t)^2 \times \frac{\pi}{4} \tag{3.27}$$

$$R_{et} = \frac{V_w \times D_{Ht}}{\mu_w} \tag{3.28}$$

Where R_{et} stands for the Reynolds value of aquatic species per month.

$$Nu_t = 2 + \left[0.4 \times (Re_t)^{1/2} + 0.06 \times (Re_t)^{2/3}\right] \times (Pr_w)^{0/4} \tag{3.29}$$

Where every month's Nusselt value of aquatic life is represented by Nu_t.

$$h_t = \frac{Nu_t \times K_w}{(2 \times r_t)} \tag{3.30}$$

Where h_t is the fish heat transfer coefficient in W/m² for each month.

$$R_{tconv} = \frac{1}{h_t \times SA_t} \tag{3.31}$$

Where heat resistance at the fish-to-water interface (K/W) is denoted by R_{tconv}.

$$R_{tcond} = \frac{\ln(r_t / r_{t1})}{2\pi \times b \times K_t} \tag{3.32}$$

Where R_{tcond} signifies heat resistance across a single fish body (K/W), r_{t1} specifies the fish's minimum radius (m), and K_t signifies the trout's thermal conductivity (average 0.5186 W/mK).

3.4 Application of Geothermal Water for Honey Processing

Raw honey requires processing instantly after the honey is extracted from the honey comb. When raw honey comes into contact with sunlight, it begins to crystallize. According to Mburu (2012), honey must be processed right away to prevent it from crystallizing. Honey processing can be divided into three categories: (1) Simple straining methods, (2) thermal heating/water bath methods, and (3) bulk processing methods. Certain theories and concepts are used in the design of processing equipment and are crucial in deciding the amount of energy required in the process, the production capacity of the equipment, the material selection for construction, and the wall thickness of the vessels and covers. Material balance, energy balance, design strain, design temperature, design stress, corrosion allowance, and joint performance are some of the factors to consider. Pressure vessels are used to describe honey processors. Any closed vessel with a diameter greater than 150 mm that is subjected to a pressure difference of more than 1 bar in relation to atmospheric pressure is classified as a pressure vessel. The description of a pressure vessel in this concept is not exact. Two kinds of vessels may be considered for design purposes: Pressure vessels with thin and thick walls. Thin-walled vessels have a wall thickness to diameter ratio of less than 1:10, whereas thick-walled vessels have a ratio greater than this (ASME, 1998). When most food processing vessels, including honey processors, are subjected to pressure load, they undergo major circumferential and longitudinal stresses in comparison to radial stress.

This is due to the fact that radial stresses are neglected during the design process. Radial stress is significant in thick-walled vessels, and circumferential stress is distributed along the wall. A number of vessels used in the food processing industry are subjected to these types of pressures, which are taken into account when designing them (Douglas, 2005). The specifications listed below must be considered in designing an integrated extractor and processor unit (Dippery and Srivastava, 2001).

3.4.1 Energy and Materials Balance

The law of mass balance was used to perform a material balance involving the volume of honey collected and processed every day. The formula for determining energy balance, which included the amount of heat needed to get a given amount of honey to the desired temperature (Suryanarayana, 2003; Holman, 1997) is Eq. (3.33):

$$q = MC\Delta T \tag{3.33}$$

Where M = mass of batch or honey, C = heat capacity of honey, ΔT = temperature difference.

3.4.2 Required Heating Medium Quantity

The energy used (calculated as $MC\Delta T$) to heat honey to target temperature was used to determine the quantity of heating medium needed to supply a given amount of heat as Eq. (3.34) (Warren et al., 2005):

$$MC\Delta T = M_1C_1\Delta T_3 \tag{3.34}$$

Where M_1 = amount of heating medium required (kg), C_1 = heat capacity of heating medium (kJ), ΔT_3 = temperature difference of heating medium.

3.4.3 Heat Flow Coefficient Calculation

Parameters like density, thermal conductivity, the heat flow coefficient, heat capacity of honey, and viscosity are used for calculation of heat transfer surface area and time needed for heating honey to pasteurize. This can be calculated as Eq. (3.35) (Nicholas, 2004).

$$\frac{uDi}{k} = a\left[\frac{L^2N\rho}{\mu}\right]^{2/3}\left[\frac{C\mu}{k}\right]^{1/3}\left[\frac{\mu b}{\mu w}\right]^{D} - 14 \tag{3.35}$$

Where U = the coefficient of heat flow, Di = designed vessel diameter (m), L agitator diameter (m), N = agitator speed revolutions per hour,

ρ = honey density (kg/m^3), μ = honey viscosity (centipoise), C = specific heat of extracted honey (kJ/kg °C), μb = bulk fluid temperature viscosity, μw = viscosity value at surface temperature condition.

3.4.4 Heat Flow Surface Area Calculation

The heat flow coefficient, the amount of heat to be transferred, and the logarithmic mean temperature difference were used to calculate the jacket surface area needed to transfer a given amount of heat from the heating medium to the honey within the vessel as shown in Eq. (3.36) (Max and Klaus, 1980).

$$Q = U\,A\Delta TL \tag{3.36}$$

Where Q = the amount of heat to be transferred, Δ = heat transfer coefficient, A = heat transfer surface area, TL = logarithmic average difference in temperature.

Where ΔTL is given as Eq. (3.37) (Holman, 1997).

$$\Delta TL = \frac{\Delta T_2 - \Delta T_1}{\ln \frac{\Delta T_2}{\Delta_{T1}}} \tag{3.37}$$

Where $\Delta T_1 = Tc_2 - Tc_1$ = lowest temperature of heating medium (Tc_2) – lowest temperature of honey (Tc_1), $\Delta T2$ = highest temperature of heating medium (Th_2) – highest temperature of honey (Th_1).

3.4.5 Calculating the Necessary Heating Time

Using the relation, the time needed to heat a given amount of honey to the desired temperature was calculated using the predetermined heat transfer coefficient, heat transfer surface area, and amount of heating medium (Eq. (3.38)) (Nicholas, 2004).

$$\ln\left(\frac{T_1 - t_1}{T_1 - t_2}\right) = \frac{W_h C_h}{Mc}\left(\frac{k_1 - 1}{k_1}\right)\theta \tag{3.38}$$

Where $k_1 = \ell uA/W_h C_h$, T_1 = medium temperature heating, t_1 = temperature of initial batch, t_2 = temperature of final batch, W_h = flow rate of medium heating (kg/h), C_h = specific heat medium heating (kJ), A = heat flow surface area (m^2), M = honey weight, Θ = time.

3.4.6 Processing

The following processes make up the honey processing stages (Yadav and Sircar, 2019).

3.4.6.1 Uncapping

We need to cover the beehive cells with wax plugs in order to keep the honey bees. The cappings are removed with a heated knife to allow the honey to flow from the beehive and be deposited in an uncapping tank.

3.4.6.2 Extractor

The frames are put in the extractor after the honey is uncapped and the honey is squeezed out. Radial and centrifugal extractors are the two most common models. Easy honey extraction is usually done with a centrifugal extractor. The extractors can be powered by electricity or by hand. The extractor's size is determined by the number of frames. The number of frames that an extractor can carry ranges from 6 to 85. The manual extractors are driven by a bicycle chain or a hand crank, while the electrical extractors are powered by a generator.

3.4.6.3 Processing

The integrated honey extractor processor assembly (Figure 3.3) includes centrifugal force extraction of honey from combs and indirect heating of honey in a jacket using a heat transfer medium (water). Using these concepts, an extractor (cylindrical vessel) is commonly constructed, consisting of honeycomb frame holders connected to the central shaft, which causes rotation and throws honey against the vessel through centrifugal force. The dimensions of the super frame are normally used to build the honey frame holders that hold the honeycomb frames during extraction (Figure 3.4).

3.5 Case Studies

3.5.1 Puga

Puga (33°13′N, 78°18′E), Leh district, Ladakh is the most promising hydrothermal system of the Indian subcontinent. It has more than 100 hot springs. The geothermally anomalous region extends for over 5 km along the Puga stream and is around 1 km wide at its widest point.

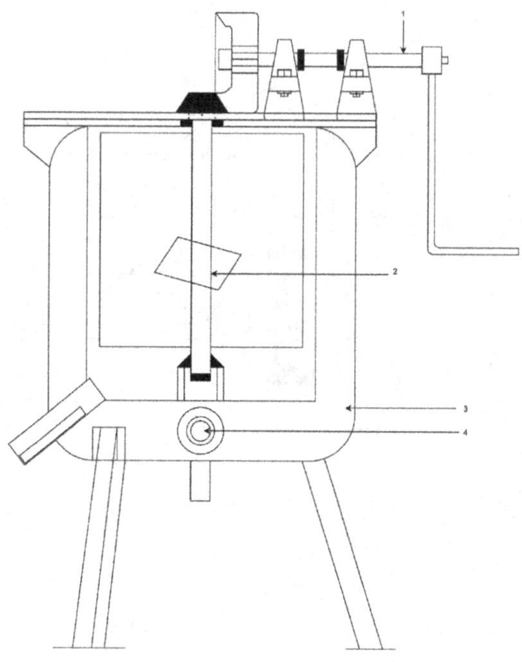

FIGURE 3.3
Integrated extractor–processor assembly for honey processing.

3.5.1.1 Methods

Six fluid samples and one mud filtrate sample were collected from seven sites of the Puga geothermal springs and stored in acid-washed high-density polyethylene (HDPE) bottles. In situ conductivity, pH, and temperature were measured with a water quality tester, TPS Aqua-TP (121136/1). To examine any elemental and isotope fractionation occurring by phase separation during cooling after sample collection, both sample aliquots that were unacidified and sample aliquots acidified with 69% HNO_3 were conditioned at each sampling site. The mud–fluid mixture sample was centrifuged in the laboratory into fluid and mud samples; the centrifuged fluid was filtered with a 0.22 mm polyethersulfone membrane. Solid samples were ground and analysed for their bulk mineralogy with a Rigaku Smartlab powder X-ray diffractometer (XRD). A ground scanning electron microscope (FE-SEM, JEOL JSM-7001F) fitted with an energy-dispersive X-ray spectrometer (EDS), Oxford Aztec X-Max and Oxford X-act, were used to examine both rock and sediment samples. Secondary ion mass spectrometry, using a modified Cameca IMF-5F, was used for in situ boron concentration analysis. After Makishima and Nakamura, inductively coupled plasma sector field mass spectrometry (ICP-SFMS) (Thermo

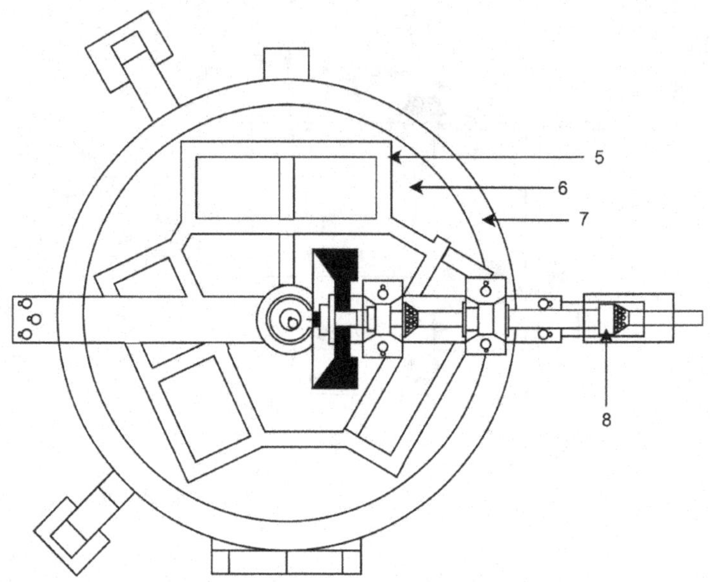

FIGURE 3.4
Number of frames to be placed in extractor–processor assembly.

Fisher Scientific Element XR) was used to evaluate major and minor elements, such as Na, Mg, Al, P, S, K, Ca, Sc, V, Cr, Mn, Fe, Co, Ni, Cu, Zn, and Ga. For boron isotope study, Nakamura et al. (1992) and Nakano and Nakamura (1998) used thermal ionization spectroscopy in static multi-collection form on a Thermo Fisher Scientific Triton TI. All these studies took place at Okayama University's Institute for Planetary Materials' Pheasant Memorial Laboratory for Geochemistry and Cosmochemistry (Nakamura et al., 2003; Nakamura et al., 2012).

Thermal releases were plotted using Na–K–Mg plots at different points. Temperatures in the reservoir were determined using an empirical equation (Giggenbach et al., 1983). To determine the fluid–mineral equilibria in the Puga geothermal region, deep fluid composition was plotted on activity–temperature plots (Shanker et al., 2000).

A specially built and engineered gradiometer, comprising fine-tuned thermistors fixed at ranges of 20, 40, 60, 80, and 100 cm from the top, was used to test shallow thermal gradients in different spots within Puga. A manually operated steel device was used to drill 1-metre holes in the ground. The gradiometer was placed into the hole and enabled to adjust to its new position. The resistances of thermistors were evaluated using a resistance bridge and a transistorized null detector after equilibrium conditions were reached. Spring geothermal water and borehole geothermal water samples were obtained and analysed for tritium levels. Chemical analysis of Puga spring waters revealed a high concentration of alkali metals like sodium, as well as a high proportion of potassium, silicon dioxide, and chloride.

3.5.1.2 Societal Benefits

Experimental space heating and refining of locally existing sulphur and borax are both done using Puga's geothermal resources. According to computer-based reservoir simulation activities, a modest 2 MW of electricity could be generated (Absar et al., 1993). The presence of iron-rich colloids, along with the formation of boron-rich, amorphous silica crusts found in the Puga geothermal system, may act as a modern-day analogue for the formation of tourmaline-bearing pyritic lamina in the c. 3.48 Ga Dresser Formation (Van Kranendonk et al., 2008; Van Kranendonk et al., 2019).

3.5.2 Manikaran

Manikaran (32°02′N, 77°21′E) is a small town in the Kullu district of Himachal Pradesh (India), which is situated on the banks of the Parbati River, a tributary of the Beas River, at an altitude of 1760 m in the north-western Himalayas.

3.5.2.1 Methods Adopted

In Manikaran, chemical tests on cold and hot water were carried out using methods recommended by the American Public Health Association (APHA, 1985). Water samples were analysed, and temperature, pH, SiO_2, Ca, Mg, Na, K, SO_4, Cl, and HCO_3 were plotted using the Langelier–Ludwig diagram (after Langelier and Ludwig, 1942). The anion variation shows a clearer distinction between the Na–HCO_3–Cl and $CaSO_4$ water classes. Fluid inclusion experiments on the Manikaran Quartzite (Sharma and Misra,1998), from which thermal releases are emitted, indicate that the trapped fluid is composed of H_2O–NaCl–KCl, since biphase aqueous inclusions have eutectic temperatures ranging from −21.1 to −23.7 °C. A graphical technique proposed by Fournier and Truesdell (1973) with a collection of enthalpy values and dissolution rate of silica (quartz) at various temperatures can be used to calculate the fraction of ground water interacting with deep thermal discharges (Alam, 2002).

The temperature, pH, electrical conductivity, alkalinity (by solution titration with 0.05N HCl), and NH_4^+ content (by ion-selective electrode) of the samples were all determined in the field. Using a LaMotte Sulphide Test Kit, the sulphide content was determined by adding three different reagents to 5 ml of solution in the following order: sulphuric acid, ferric chloride hexahydrate, and ammonium phosphate, followed by a colour scale. For the main finite element method, all samples were filtered through 0.45 m cellulose filters and acidified with HCl 6 M and HNO_3 4 M for the minor trace element method. The samples were held in HDPE bottles and sterile polypropylene tubes for the analysis of major and minor trace elements. Ion chromatography was used to determine significant cations and anions ($Ca2^+$, $Mg2^+$, Na^+, K^+, F^-, Cl^-, Br^-, $SO2^-$, NO^-), while ICP-MS was used to determine minor (SiO_2, Al) and trace elements (B, Sb, Li, Sr). Atomic emission spectrometry was used to determine the iron content, while hydride generation atomic absorption spectrometry (HGAAS) was used to determine the arsenic concentrations. Capasso and Inguaggiato (1998) collected dissolved gases, which were then examined using a gas chromatograph equipped with two well-known sensors (FID – flame ionization sensor and TCD – thermal conductivity sensor) and N_2 and H_2 as carrier gases, respectively. The total dissolved inorganic carbon (TDIC) and pCO_2 of the investigated waters were determined using the PHREEQC code v. 2.12 and the Lawrence Livermore National Laboratory (LLNL) database, with the groundwater chemical composition, outlet temperature, and pH as input data.

3.5.2.2 Societal Benefits

Agricultural production is a direct utilization of geothermal energy; the region's geothermal releases are channelled and used for space heating in homes, schools, hotels, and tourist complexes; they are also used for the operation of cold storage facilities for the fruit industry, food processing, and greenhouse farming. Indirect usage is for regional power production, which can provide a consistent supply to the village at a lower cost for large-scale production of sheep farms, wool manufacturing, animal husbandry, poultry farming, fish farming, and carpet shawl manufacturing, all of which provide jobs for the local community.

References

1. Absar, A., Jangi, B.L., Pandy, S.N., Srivastava, G.C., Singh, J.R., 1993, Geological and Volcanological studies at Barren Island Volcano, Andaman and Nicobar Islands, *Record, GSI*, Vol. 126(8), pp. 218–219.

2. Alam, M.A., 2002, Hydrogeochemistry of Thermal Springs in Manikaran, Kullu District, Himachal Pradesh, India. MSc Dissertation Report, Department of Earth Sciences, Indian Institute of Technology, Bombay, India, 2002, unpublished.
3. Antoine-Frostburg State University, 2016, Water Density Calculator. Available at: https://antoine.frostburg.edu/chem/senese/javascript/water-density.html.
4. APHA, *Standard Methods for Examination of Water and Wastewater*, 1985, APHA, AWWA, WPCF (16th Ed), 199, 210, 228, 237, 246, 269, 287, 467.
5. ASME (American Society of Mechanical Engineers), 1998, *Boiler and Pressure Vessel Code*, pp. 280–335.
6. Bandyopadhyay, U.K., Santha Kumar, M.V., 2013, Studies on Biology of Whitefly, *Dialeuropora decempuncta* on Mulberry, *Annals of Plant Protection Science*, Vol. 16 (2), pp. 498–500.
7. Ben Mohammed, M., 2015, Geothermal Energy Development: The Tunisian Experience, In *Proceedings, World Geothermal Congress*, Melbourne, Australia, 8 p.
8. Bergman, T.L., Lavine, A.S., Incropera, F.P., Dewitt, D.P., 2011, *Fundamentals of Heat and Mass Transfer*, John Wiley.
9. Brautigam, F.C., 2015, Contribution of Fall-Stocked Brown Trout to a Winter Fishery: A Comparison of Trout Stocked in October and December. Fishery Final Report Series No. 15-1. Maine Department of Inland Fisheries and Wildlife Fisheries and Hatcheries Division.
10. Cameron Morton, A., 2013, Assessing the Performance of a Reservoir-Based Water Source Heat Pump. MSc thesis, University of Strathclyde Engineering, Department of Mechanical and Aerospace Engineering.
11. Capasso, G., Inguaggiato, S., 1998, A Simple Method for the Determination of Dissolved Gases in Natural Waters. An Application to Thermal Waters from Vulcano Island, *Applied Geochemistry*, Vol. 13 (5), pp. 631–642.
12. Carew, A.L., Mitchell, C.A., 2008, Teaching Sustainability as a Contested Concept, *Journal of Cleaner Production*, Vol. 16, pp. 105–115.
13. Chandrasekharam, D., Chandrasekhar, V., 2020, Geothermal Energy Resources of India: Country Update, In *Proceedings, World Geothermal Congress 2020*, Reykjavik, Iceland, 13 p.
14. Chappell, M.J., Wittman, H., Bacon, C.M., Ferguson, B.G., Barrios, L.G., Barrios, R.G., Jaffee, D., Lima, J., Méndez, V.E., Morales, H., Soto-Pinto, L., Vandermeer, J., Perfecto, I., 2013, Food Sovereignty: An Alternative Paradigm for Poverty Reduction and Biodiversity Conservation in Latin America, *F1000Research*, Vol. 2, p. 235.
15. Chiodi, A.L., Filipovich, R.E., Esteban, C.L., Pesce, A.H., Stefanini, V.A., 2020., Geothermal Country Update of Argentina: 2015–2020, In *Proceedings, World Geothermal Congress 2020*, Reykjavik, Iceland, 13 p.
16. Darma, S., Yaumil, I.L., Naufal, M., Shidqi, A., Riyanto, T.D., Daud, M.Y., 2020, Country Update: The Fast Growth of Geothermal Energy Development in Indonesia, In *Proceedings, World Geothermal Congress 2020*, Reykjavik, Iceland, 9 p.

17. Dippery, Jr, D., Srivastava, T., 2001, Pressure Vessel Design Using Boundary Element Method with Optimization, *Transactions on Modelling and Simulation*, Vol. 27, pp. 267–270.
18. Douglas, W., 2005, *Pressure Vessel Design DANOTES*, McGraw-Hill. p. 10–34.
19. Engel, W., Hoppe, U., 1987, Aqueous hair preparations containing sericin and pelargonic acids, Ger Offen DE 3603595 A1 (to Beiersdorf A. G, Ger) 13 August 1987, P 4, *Chemical Abstracts*, Vol. 108 (16), p. 137689.
20. Fournier, R.O., Truesdell, A.H., 1973, An Empirical Na-K-Ca Geothermometer for Natural Waters, *Geochimica et Cosmochimica Acta*, Vol. 37, pp. 1255–1275.
21. Fričovský, B., Černák, R., Marcin, D., Blanárová, V., Benková, K., Pelech, O., Fordinál, K., Bodiš, D., Fendek, M., 2020, Geothermal Energy Use – Country Update for Slovakia, In *Proceedings, World Geothermal Congress 2020*, Reykjavik, Iceland, 19 p.
22. Garg, P., Gary, H.P., 2000, *Solar Energy: Fundamentals and Applications*, Tata McGraw-Hill Education, p. 434.
23. Giggenbach, W.F., Gonfiantini, R., Jangi, B.L., Truesdell, A.H., 1983, Isotopic and Chemical Composition of Parbati Valley Geothermal Discharges, North-west Himalaya, India, *Geothermics*, Vol. 12, pp. 199–222.
24. Good, I.L., Kenoyer, J.M., Meadow, R.H., 2009, New Evidence for Early Silk in the Indus Civilization, *Archaeometry*, Vol. 50, pp. 457.
25. Hamza's Reef, 2016, Water Properties Calculator. Available at: www.hamzasreef.com.
26. Holman, J.P., 1997, *Heat Transfer*, McGraw–Hill Companies, New York, pp. 560–574.
27. Hopwood, B., Mellor, M., O'Brien, G., 2005, Sustainable Development: Mapping Different Approaches, *Sustainable Development*, Vol. 13, pp. 38–52.
28. Kępińska, B., 2020, Geothermal Energy Country Update Report from Poland, 2015–2019, In *Proceedings, World Geothermal Congress 2020*, Reykjavik, Iceland, 13 p.
29. Langelier, W.F., Ludwig, H.F., 1942, Graphical Method for Indicating the Mineral Character of Natural Waters, *Journal of the American Water Works Association*, Vol. 34, pp. 335–352.
30. Lund, J.W., 2010, Direct Utilization of Geothermal Energy, *Energies*, p. 7.
31. Maji, M.D., Rao, K.V.S.N, Das, C., 2008, Response of Some Elite Mulberry Varieties to Foliar Diseases under Koraput Condition, *Bulletin of Indian Academy of Sericulture*, Vol. 12 (2), pp. 35–41.
32. Mburu, M., 2012, Cascaded Use of Geothermal Energy: Eburru Case Study, *Geo-Heat Center, Quarterly Bulletin*, Vol. 30, pp. 21–26.
33. Microelectronics Heat Transfer Laboratory (MHTL), 2016, Fluid Properties Calculator. Available at: www.mhtl.uwaterloo.ca/old/onlinetools/airprop/airprop.html.
34. Miyashita, T., Sweat and sebum absorbing cosmetics containing cellulose fibres, Jpn Kokai Tokkyo Koho, Jap 11152206 A2 (to Hsan Sangyo K Japan) 8 June 1999, Heisei, P 3, *Chemical Abstracts*, Vol. 131 (2), p. 23271.
35. Morozov, Y., Barylo, A., Lysak, O., 2020, Geothermal Energy Country Update Report from Ukraine, 2020, In *Proceedings, World Geothermal Congress 2020*, Reykjavik, Iceland, 6 p.

36. Murray, D., 2014, Effect of Biomass in Recirculation Aquaculture Water Heating and Cooling Systems, MOFS Thesis, Institute of Harford.
37. Nakamura, E., Ishikawa, T., Birck, J.L., Allegre, C.J., 1992, Precise B isotopic analysis of natural rock samples using a B-mannitol complex, *Chemical Geology*, Vol. 94, pp. 193–204.
38. Nakamura, E., Makishima, A., Moriguti, T., Kobayashi, K., Sakaguchi, C., Yokoyama, T., Tanaka, R., Takeshi, K., Takei, H., 2003, Comprehensive Geochemical Analyses of Small Amounts (<100 mg) of Extra-terrestrial Samples for the Analytical Competition Related to the Sample Return Mission MUSES-C, In *The Institute of Space and Astronautical Science Report*, SP 16, Japan Aerospace Exploration Agency, Tokyo, pp. 49–101.
39. Nakamura, E., Makishima, A., Moriguti, T., Kobayashi, K., Tanaka, R., Kunihiro, T., Tsujimori, T., Sakaguchi, C., Kitagawa, H., Ota, T., Yachi, Y., Yada, T., Abe, M., Fujimura, A., Ueno, M., Mukai, T., Yoshikawa, M., Kawaguchi, J., 2012, Space Environment of an Asteroid Preserved on Micrograins Returned by the Hayabusa Spacecraft, *Proceedings of the National Academy of Science of the United States of America*, Vol. 101, pp. E624–E629.
40. Nakano, T., Nakamura, E., 1998, Static multicollection of Cs2BO2+ ions for precise B isotope analysis with positive thermal ionization mass spectrometry, *International Journal of Mass Spectrometry and Ion Processes*, Vol. 176, pp. 13–22.
41. Nicholas, P.C., 2004, *Handbook of Chemical Engineering Calculations*, McGraw-Hill, London, pp. 7.1–7.38.
42. Peters, M.S., and Timmerhaus, K.D., 1980, *Plant Design and Economics for Chemical Engineers*, McGraw-Hill International Book Company, London, pp. 12–57, 632–660.
43. Ragnarsson, A., Steingrímsson, B., Thorhallsson, S., 2020, Geothermal Development in Iceland 2015–2019, In *Proceedings, World Geothermal Congress 2020*, Reykjavik, Iceland, 14 p.
44. Raje, S.S., Rekha, V.D., 1998, Man-made Text India, Vol. 41 (6), pp. 249–254.
45. Taylor, R., 2005, Identifying New Opportunities for Direct Use Geothermal Development, California Energy Commission, California, Consultant report.
46. Saeedi, M., Moradi, M., Hosseini, M., Emamifar, A., Ghadimi, N., 2019, Robust Optimization Based Optimal Chiller Loading under Cooling Demand Uncertainty, *Applied Thermal Engineering*, Vol. 148, pp. 1081–1091.
47. Shanker, R., Absar, A., Srinvastava, G.C., Bahpai, I.P., 2000, *A Case Study of Puga Geothermal System, Ladakh, India*, Geological Survey of India, Lucknow, India.
48. Sharma, R., Misra, D.K., 1998, Evolution and Resetting of the Fluids in Manikaran Quartzite, Himachal Pradesh: Applications to Burial and Recrystallization, *Journal of the Geological Society of India*, Vol. 51, pp. 785–792.
49. Suryanarayana, O.A., 2003, Design and Simulations of Thermal Systems, McGraw-Hill Vol. 10 (55), p. 229–250.
50. Tian, T., Zhang, W., Wei, J., Jin, H., 2020, Rapid Development of China's Geothermal Industry – China National Report of the 2020 World Geothermal Conference, In *Proceedings, World Geothermal Congress, 2020*, Reykjavik, Iceland, 9 p.

51. Van Kranendonk, M.J., Philippot, P., Lepot, K., Bodorkos, S., Pirajno, F., 2008, Geological Setting of Earth's Oldest Fossils in the ca. 3.5 Ga Dresser Formation, Pilbara Craton, Western Australia, *Precambrian Research*, Vol. 167, pp. 93–124.
52. Van Kranendonk, M.J., Djokic, T., Poole, G., Tadbiri, S., Steller, L., Baumgartner, R., 2019, Depositional Setting of the Fossiliferous, c. 3480 Ma Dresser Formation, Pilbara Craton: A Review, In *Earth's Oldest Rocks*, edited by M.J. Van Kranendonk, V.C. Bennett, and E.J. Hoffmann, 2nd ed., Elsevier, Amsterdam, pp. 985–1006.
53. Wakana, F., 2020, Geothermal Status in Burundi, In *Proceedings, World Geothermal Congress 2020*, Reykjavik, Iceland, 6 p.
54. Warren, L., McCabe, J., Julian, S.C., Harriot, P., 2005, *Unit Operations of Chemical Engineering*, 7th ed., McGraw, New York, pp. 253–285, 444–514.
55. Yadav, K., Sircar, A., 2019, Application of Low Enthalpy Geothermal Fluid for Space Heating and Cooling, Honey Processing and Milk Pasteurization, *Case Studies in Thermal Engineering*, Vol. 14, pp. 100499.
56. Yamada, H., Yamasaki, K., Zozaki, K., 2001, Nail cosmetics containing sericin, PCT Int Appl WO 2001015660 A1 (to Teikoku Seiyaku Co Ltd Seiren Co Ltd Japan) 8 March 2001, P 15, *Chemical Abstracts*, Vol. 134 (14), pp. 197888.
57. Yasukawa, K., Nishikawa, N., Sasada, M., Okumura, T., 2020, Country Update of Japan, In *Proceedings, World Geothermal Congress 2020*, Reykjavik, Iceland, 7 p.
58. Yoshioka, M., Segawa, A., Veda, A., Omi, S., 2001, UV absorbing compositions containing fine capsules, Jpn Kokai Tokkyo Koho Jap2001049233 A2 (to Seiwa Kasei K K Japan) 20 February 2001, P 14, *Chemical Abstracts*, Vol. 134 (14), p. 197870.

4

Indirect Applications: Application of Geothermal Water for Milk Pasteurization, Cloth and Food Drying, Desalination, and Space Heating and Cooling

4.1 Indirect Applications of Geothermal Water around the World

Geothermal heating systems are the most widely used geothermal technology, responsible for 71.6% of generation capacity and 59.2% of annual energy consumption worldwide. In the heating phase, the generating capacity is 77,547 MWt, and the power use is 599,981 TJ/yr, with a capacity factor of 0.245 (Ait Ouali et al., 2020). The number of nations with installations grew from 26 in 2000 to 33 in 2005, 43 in 2010, 48 in 2015, and 54 in 2020, despite the fact that the majority of the installations are in North America, Europe, and China (Tian et al., 2020; Gondwe et al., 2020). The corresponding number of connected 12-kW units (average of households in the United States and Western Europe) is 6.46 million. This represents a 54% rise over the number of units recorded in 2015, and more than twice the number documented in 2010. Individual groups, on the other hand, vary in size from 5.5 kW for household usage to massive units of over 150 kW for commercial and industrial use (Yasukawa et al., 2020). Apart from the northern regions, many systems in the United States are built for major cooling demand and enlarged for heating; hence, they are expected to generate only 2000 equivalent full-load heating hours annually (capacity factor of 0.23). Several systems in Europe are designed for heat supply and are frequently intended to offer base load with fossil energy peaking, although a few of these systems, such as those in the Nordic nations, may operate for up to 3000 quantity full-load warming hours yearly (capacity index of 0.34), mainly in Finland (Kallin, 2019). A figure of 2200 h/year, and greater for some of the Nordic regions, has been used for energy output (TJ/yr) except if the exact number of equivalent full-load warming hours was supplied. Compared with an average coefficient of performance

DOI: 10.1201/9781003204671-4

(COP) of 3.5, which enables 1 unit of electricity input (typically electricity) to 2.5 units of electricity consumption for a geothermal component of 71% of the rated capacity (i.e. (COP – 1)/COP = 0.71), the energy use recorded for the heat pumps was calculated from the installed capacity. Because heat is dumped into the surface or groundwater, the cooling was not regarded as geothermal in this situation (Goldbrunner, 2020).

Cooling, on the other hand, has a place in the replacement of fossil fuels and the decrease of greenhouse gas emissions and is thus further discussed later. China, the United States, Sweden, Germany, and Finland contribute 77.4% of the entire implementation (MWt), and 83.5% of power generated (TJ/yr).

Since 2015, the capacity factor of space heating, comprising individual space heating and district heating, climbed by 68.0%, while the annual energy demand climbed by 83.8%. The entire installed capacity is presently 12,768 MWt with a total electricity use of 162,979 TJ. District heating, on the other hand, accounts for 91% of power capacity and 91% of annual energy demand (Goldbrunner, 2020).

Iceland, China, Germany, Turkey, and France lead the district heating industry in terms of efficiency and yearly energy usage, while Japan, the United States, Hungary, Russia, and Turkey lead the individual space heating segment in terms of installation (MWt). Japan, the United States, Russia, Turkey, and Switzerland are the top five nations in terms of yearly energy usage (TJ/yr), with a total of 29 nations (Yasukawa et al., 2020). These five representatives account for roughly 90% of global district heating usage and approximately 75% of global consumer space heating.

Industrial process heating is usually massive, consuming a lot of resources, and runs all year. Examples include concrete curing (Guatemala and Slovenia), bottling of water and carbonated drinks (Bulgaria, Serbia, and the United States), milk pasteurization (Romania and New Zealand), leather industry (Serbia and Slovenia), chemical extraction (Bulgaria, Poland, and Russia), CO_2 extraction (Iceland and Turkey), pulp and paper processing (New Zealand), iodine and salt extraction (Vietnam), and borate and boric acid production (Italy).

In comparison to 2015, the capacity factor is 852 MWt and the energy consumption is 16,390 TJ/yr: a 38.8% and 56.8% rise, respectively. Heat for production plants has one of the largest capability factors of all direct utilization, at 0.610, up from 0.540 in 2015 but lower than 0.699 in 2010 and 0.712 in 2005, as predicted, due to almost year-round activity. The reduction in capacity factor is not explained, but it may be attributed to more productive processes and energy usage as well as fewer operating hours annually. Iceland, Hungary, Russia, New Zealand, and China are the top energy users (in TJ/yr), responsible for 98% of the total (Daysh et al., 2020).

The total capacity and average energy usage of geothermal resources for greenhouse and ground heating increased by 24% and 23%, respectively, globally. The capacity factor is 2459 MWt with an annual energy consumption of 35,826 TJ. Hungary, the Netherlands, China, and Turkey are the top nations in annual energy consumption (TJ/yr), responsible for around 83% of the global total (Bakema et al., 2020). Many nations do not differentiate among greenhouse and open-air ground heating, and only a few have recorded the total area heated. Vegetables and herbs are the most common major crops grown in greenhouses, although tree saplings, cacti, and fish in lakes (in the United States), as well as fruits like bananas (in Iceland), are also cultivated. Developing nations have a comparative advantage over more industrialized countries in this industry because labour is one of the main costs. Taking 20 TJ/yr/ha as the total energy demand for greenhouse warming from 2000, the 35,826 TJ/yr refers to around 1791 ha of greenhouses heated globally, a 23.4% rise from 2015.

4.2 Application of Geothermal Water for Milk Pasteurization

Milk is a nutrient-dense food that is an integral part of pastoralist societies' diets. Milk quality degrades quickly after collection due to enzymatic activity and microbial growth, especially in unsanitary generation and processing environments at room temperature. To avoid enzyme reactions and growth of microorganisms, milk must be handled at high temperatures, such as by pasteurization or the ultra-high temperature (UHT) procedure (Perko, 2011; Torkar and Golc Teger, 2008).

Milk pasteurization and processing can be done with geothermal energy, whereas milk evaporation and the UHT method can be done with geothermal heat. Figure 4.1 shows a flow diagram of the pasteurization of milk. The warm milk from the homogenizer preheats fresh chilled milk at a temperature of 3 °C in a plate-type heat exchanger up to 71 °C. The pasteurized milk is then transferred through a geothermal plate-type heat exchanger, which heats it to a minimum of 78 °C for 15 seconds. The warm milk is transferred to the homogenizer after pasteurization and then to the plate heat exchanger, in which it is chilled to 12 °C. Until processing and packaging, it is cooled to 3 °C using cold water in a plate-type heat exchanger. The input temperature of geothermal warm water is around 87 °C, while the outlet temperature is around 77 °C (Lund, 1995). The warm water and milk fluid will flow via the plate-type heat exchanger in opposite ways on both sides of the plates, as shown in Figure 4.1. The positioning of the plate seals, which are mounted to avoid contamination of the milk and warm water, controls the water movement and distribution.

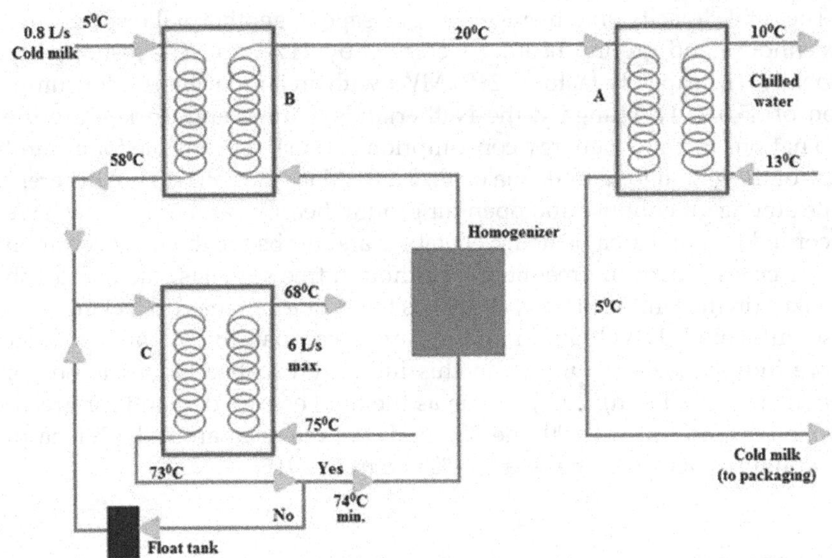

FIGURE 4.1
Flow diagram of milk pasteurization unit.

4.3 Application of Geothermal Water for Cloth and Food Drying

4.3.1 Food Drying

Tomatoes and other veggies can be preserved in a variety of ways. The amount of time it takes to dry a tomato is based on a range of factors, including the tomato species, the soluble solids content of the new products, the air humidity, the size of the tomato sections, the air temperature and velocity, and the drying service's effectiveness (Fytikas et al., 2000). The frequency of drying is the most important component in determining the finished product's quality (Lund, 2000). Pre-drying procedures (such as size selection, cleaning, and tray placement), drying or dehydration, and post-dehydration procedures (such as assessment, screening, and packing) are the procedures that dried tomatoes go through in general (Mujundar, 1987).

Sun drying has the benefits of ease and low initial expenditure, but it takes a long time. Long drying times have a negative impact on product quality: Dirt and bugs can compromise the finished product, and enzymatic and microbial growth can occur. But, on the other hand, commercial drying at extreme temperatures (90 °C) results in colour and scent loss, as well as case stiffening (the production of a hard outer shell) that prevents the inside of the object from drying. The best temperature for drying tomatoes is between 45 and 55 °C, which allows the dried tomatoes to keep their nutrients (such as vitamins and lycopene, the nutrient that accounts for the deep red colour of tomatoes) and flavours (Shi et al., 1999).

As shown in Figure 4.2, the entire tomato dehydration procedure can be separated into three phases: pre-drying treatment, drying, and post-drying processing. The unprocessed tomatoes (Roma type, possibly one of the most ideal types for drying) are prepared for the dehydration treatment using a pre-drying procedure.

This process begins with a choice of tomatoes based on their ripeness and quality. Approximately 50–70% of the tomatoes are chosen. The tomatoes are separated into two size ranges: those weighing more than 90 grammes and those weighing less. The fresh tomatoes are then stored in containers, washed with water to remove dust, dirt, plant pieces, and other debris, and then sliced into two halves and put in steel platters (mesh type, 100×50 cm^2). It should be emphasized that fresh tomatoes do not need blanching due to their high antioxidant content. A trench dryer is used to finish the drying process. The following are the main features of this dryer (Figure 4.2):

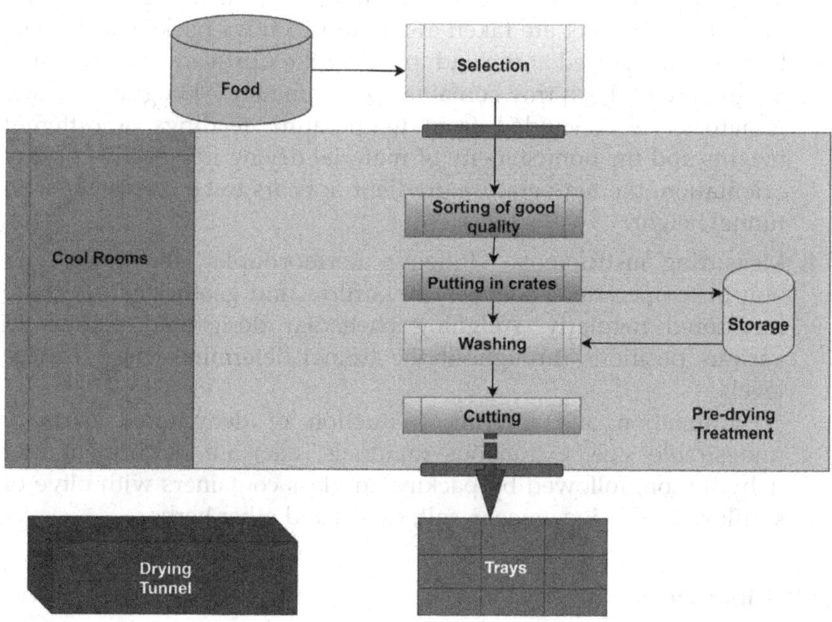

FIGURE 4.2
Flow diagram of food drying unit.

1. With a capability of 300,000 kcal, a finned-coil heat exchanger is used to heat the drying air. The "cold" air enters the pump at ambient temperatures (20–35 °C) and exits at a nearly specific temperature of 55 °C. The heat of the entering geothermal water is 59 °C, whereas the water temperature at the outflow is 51–53 °C. In the first drying stage, the average flow rate of water is around 25 m^3/h. The total dissolved solids (TDS) of geothermal water is extremely low, and it will not produce scaling or oxidation (Andritos et al., 2003).
2. The system comprises two centrifugal fan devices with a combined rated output of 7 kW. Only a tiny portion (30%) of this electricity could be used with the help of an inverter and during the drying device's operations in 2001. The pipeline's flow rate was 10,000–12,000 m^3/h, and the tunnel's surface air velocity (without the item pans) was 1.7 m/s. Air velocity rises by 20–50% in the vicinity of laden trays, which partly obstruct the tunnel's cross-section, based on the area within the tunnel (Lund, 2000).
3. Polyurethane aluminium plates make up the 14-meter-long rectangular tunnel (width 1 m, height 2 m). The hot air travels in the opposite direction as the trays pass through the tunnel. The tomato-loaded platters are mounted at the tunnel's entrance and transported in a semi-continuous fashion to the tunnel's exit (where the warm air reaches the tunnel): every 45 minutes, in sequence, 25 trays with dried food product are taken away, and 25 trays packed with fresh tomatoes are placed at the end, pushing the upstream trays towards the other end. Each tray contains approximately 7 kilograms of raw tomatoes. As concluded from temperature readings at different heights and the homogeneity of material drying irrespective of tray orientation, the temperature gradient appears to be consistent with tunnel height.
4. Measuring instruments. Utilizing thermocouples, the intake and output temperatures between the airflow and geothermal fluid are monitored regularly. Weighing particular designated platters at various positions throughout the tunnel determines the moisture levels.

   Examination and filtering (reduction of dehydrated parts of undesirable size, extraneous materials, etc.) are performed after dehydration, followed by packing in glass containers with olive or sunflower oil, wine vinegar, salt, garlic, and other herbs.

4.3.2 Cloth Dryer

The geothermal vapour is utilized to heat the wash machines directly. The application of geothermal steam, which is readily accessible in the

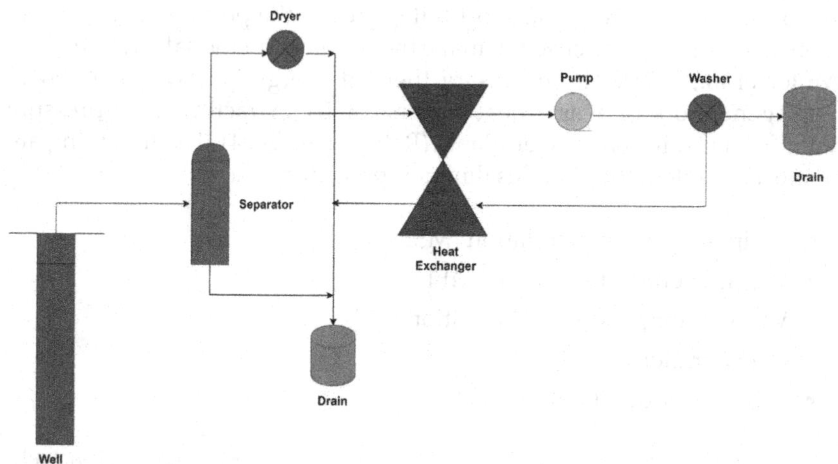

FIGURE 4.3
Flow diagram of cloth drying unit using geothermal water.

neighbourhood, results in significant electricity savings (Figure 4.3). A heat exchanger uses a portion of the geothermal vapour to warm cold water to 90 °C to be used in washing machines. Discharge water from the dryers, heat exchanger, and washing machines is pumped into a cooling tank before being released back into the environment (Arason, 2003). The laundry uses around 0.2–0.3 kg/s of geothermal vapour at full load. Because of the use of vapour for warming, the laundry can conserve the electricity that would otherwise be used for heating and drying. In addition to utilizing geothermal vapour, the laundry only utilizes eco sustainable detergents, reducing the laundry's environmental impact (Líndal, 1992).

4.4 Application of Geothermal Water for Desalination

There are two sorts of technology that are widely used and economically proven: Thermal reactions involving phase transition and membrane filtration, both comprised of various procedures. Furthermore, other technologies, such as chilling and ion exchange, are not frequently used. To generate fresh water, all of them need either traditional or renewable energy.

Desalination, commonly known as distillation, is among the oldest methods of converting seawater and brackish water into drinkable water. The concepts of boiling, evaporation, and condensation are basics of this phenomena. The water is heated till it comes to the point of evaporation, where the steam is condensed to make fresh water and the salt is left behind (Winter et al., 2005). The necessary thermal energy is now generated in steam generators or waste heat boilers, or by extracting back-pressure steam from turbines in power plants (Raluy et al., 2004). The following are the most prevalent thermal desalination procedures:

- Multi-stage flash distillation (MSF),
- Multiple-effect distillation (MED),
- Vapour-compression evaporation (VC),
- Cogeneration,
- Solar water desalination.

Figure 4.4 shows a detailed classification of different desalination technologies of major processes and alternative processes. The major processes

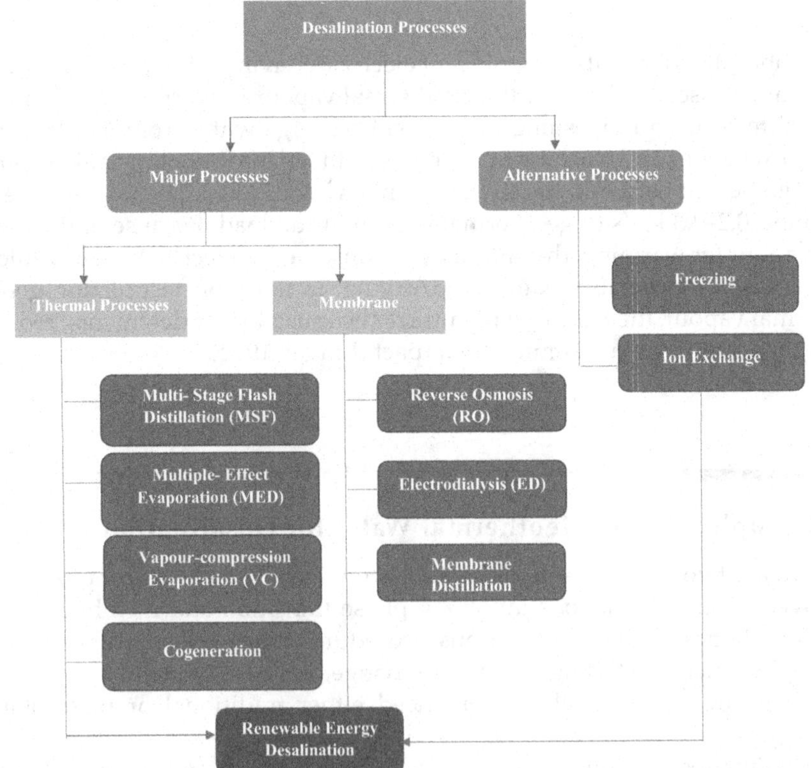

FIGURE 4.4
Flow diagram for classification of water desalination technologies.

are further classified into thermal and membrane processes, while alternative processes are further divided into freezing and ion exchange methods.

4.4.1 Multi-stage Flash Distillation

For almost a century, water distillation in a vessel running at a lowered pressure has been employed to provide a lower boiling point for water. The MSF method was invented in the 1950s by Weirs of Cathcart in Scotland, and it saw major development and potential availability throughout the 1960s due to its financial scale and ability to run on low-grade steam (Hamed et al., 2001). MSF presently produces roughly 64% of the entire global desalination plants' water production. The Arab world is home to the majority of MSF plants. Despite the fact that the MSF procedure is the most efficient method for producing fresh water from seawater, it is a high-energy process that necessitates both thermal and mechanical power (Hamed et al., 2001).

The MSF procedure involves heating supply water (brine water) in a container called a brine warmer to a temperature well below its boiling point. The warmed seawater is pumped into a number of tanks in order, where the reduced ambient pressure causes the fluid to boil and evaporate quickly. Since the water virtually flashes into vapour, this quick entry of boiling water into the low-temperature chamber is known as the "flashing" phenomenon (Water Desalination Technologies, 2001). As boiling persists till the water condenses and vaporization ceases, only a small fraction of this water is transformed into water vapour. The percentage is primarily determined by the pressure within the phase. By condensing the vapour steam formed by flashing on the pipes of heat exchangers which pass over each step, the vapour steam is turned into fresh water. The tubes are kept cool by the input feed water that goes to the heat exchanger. This, in return, warms the input water and improves thermal efficiency by lowering the amount of heat energy used in the heat exchanger to increase the seawater's temperature level. The input water that was moved from one step to the next was boiled continuously without applying further heat in an MSF system that used a sequence of steps set at successively lower atmospheric pressures. In most cases, an MSF plant has between 15 and 25 phases (Buros, 2000). MSF technology produces 84% of the global desalination capacity, that is created via distillation operations. The majority of MSF plants were developed in the Middle East, where energy supplies are plentiful and cheap.

4.4.2 Multi-effect Distillation

For water treatment, the MED technique is the original large-scale distillation technology. MED plants currently produce around 3.5% of the world's desalted water (Wagnick, 2000). Its most prominent features are high quality of distilled water, high production capacity, and high heat effectiveness (Wagnick, 2000; Al-Shammiri et al., 1999). Furthermore, MED has long been utilized in the commercial distillation industry for the evaporation of sugarcane juice in sugar production and for the evaporation of salt in the manufacture of salt utilizing the evaporative technique. The MED procedure, like MSF, occurs in a succession of containers or evaporators known as impacts, and it likewise employs the evaporation and condensation principles by lowering ambient pressure in the different effects. This method allows the seawater supply to boil numerous times without the need for extra heat after the first effect. After being preheated in pipes, seawater reaches the first action and is elevated to boiling point. To enhance quick evaporation, saltwater is poured onto the top of the evaporator pipes. External energy in the form of steam, typically from a dual-purpose power plant, heats the evaporator pipes. The steam condenses at the other end of the pipes, and the condensed steam is returned to the power plant for boiler water. The steam efficiency of the MED plant is related to the number of impacts. The overall number of impacts is restricted by the total temperature range accessible as well as the minimal temperature gradient between impacts. In the first impact, only a fraction of the seawater poured into the tubes evaporates. The remainder of the feed water is directed to the second reaction, where it is fed into a circular tube once more. The vapours formed in the first action heat these pipes, which in turn heat the pipes. From impact to impact, the evaporator and condenser processes are repeated, each at a decreasing pressure and temperature. This normally persists for 4–21 effects, with a performance ratio of 10 to 18 in large plants (Khawaji et al., 2008).

4.4.3 Vapour-compression Evaporation

Other processes, such as MED and single-effect vapour compression, are employed in conjunction with the vapour-compression extraction process. The energy for evaporating the seawater originates from the compression of vapour in this operation. The concept of lowering the boiling point by lowering the pressure is used in VC systems. A motorized compressor (mechanical vapour compression) and a steam jet (thermal vapour compression) are employed to condense the liquid water and provide enough heat to evaporate the entering seawater. The motorized compressor is often powered by electricity or diesel. To increase the quantity of energy and evaporate the saltwater, VC machines have been designed in a variety of layouts. The compressor leaves a void in the evaporator before compressing and condensing the vapour extracted from it inside a circular tube. Additional vapour is produced when saltwater is poured on the exterior of the hot tube bundle, where it boils and partly evaporates. A venturi aperture at the steam jet generates and absorbs vapour from the evaporation, resulting in reduced ambient pressure with the steam-jet kind of VC distillation machine, known as a thermo compressor. The steam jet compresses the recovered water vapour. This combination condenses on the pipe wall, providing heat energy (condensation heat) to evaporate the saltwater on either side of the pipe wall.

4.4.4 Cogeneration System for Power and Water Desalination

Energy can be used in dual-use or cogeneration technologies, in which the sources of energy can fulfil several activities, such as electricity generation and desalination. Cogeneration plants linked to multi-stage flash seawater desalination and functioning on saltwater generate the majority of desalinated drinkable water and electricity in the Arabian Gulf nations and North Africa (Buros, 2000). Alternative distillation methods, such as heat vapour compression and MED, are beginning to gain traction in the market, but the MSF method remains the workhorse of the desalination business. Over nearly half a century of plant planning and installation, this technology has shown its dependability and versatility. The MSF method is the only economically viable option for large-scale desalination capacity. On the other hand, for cogeneration, there are several commercially possible choices for providing the requisite electrical power and vapour for desalination (El-Nashar, 2001). High-pressure vapour is used to power turbines in energy generation, with the steam generated by boilers at temperatures as high as 540 °C. The thermal energy level of this vapor decreases. As noted earlier, distillation facilities require vapour with temperatures below 120 °C, which may be easily accessed at the turbine's output after most of its power has been used to generate electricity. The steam is used in seawater desalination, and the condensation is transferred to the boiler to be warmed before being utilized in the turbine. The fundamental advantage of this approach is that it requires much less fuel than individual plants, and energy costs are an important consideration in any seawater desalination. The continuous linkage between the water treatment plant as well as the power station, on the other hand, can create problems in wastewater treatment when the need for electricity is reduced or when the rotor or generator is down for maintenance. The scale of the water treatment plant can be effectively combined with power cogeneration, ensuring that the proportion of desalted water to power output meets the water and power needs of the community at large. Power prices have also been reduced thanks to cogeneration units. Other kinds of energy generation, on the other hand, have reduced costs by utilizing water heaters on gas turbines, water heaters, and other industrial activities like solid waste incineration (Buros, 1999).

4.5 Application of Geothermal Water for Space Heating and Cooling

4.5.1 Ground Source Heat Pump

Another very efficient, clean, and long-lasting geothermal ground source heat pump technology for space conditioning is the thermal ground source heat pump (GSHP) system. A subsurface source or geo-exchange technology, comprising a heat pump coupled to a sequence of subterranean pipes, is a traditional technology that could save up to 50% of power (Phetteplace, 2007). In a single system, a heat pump is utilized for air conditioning. The tubes can be installed in lateral tunnels underneath the surface of the ground or in vertical wellbores several hundred feet underground. To transfer heat from one location to another, a heat-conveying liquid, usually water, is pumped via tubes. The water heater can transport thermal energy from the subsurface to the house if the underground temperature of the water is higher than the ambient temperature. The heat exchanger can also work the other way, transporting heat from a house's surrounding atmosphere to the ground and chilling the house. The heating and air conditioning processes in GSHPs only use a minimal amount of power. The heat exchanger can provide up to five times the heat from the subsurface for each and every unit of energy consumed to control the equipment, leading to a net energy profit.

4.5.2 Geothermal Space Cooling

A technique wherein chilled liquid is delivered through tubes from a centralized air conditioning plant to houses for cooling systems is known as space conditioning (Kaushik and Nand, 2015). The three basic components of a space air conditioning system are the cooling source, distribution network, and consumer placements (also known as energy treatment plants (ETS)).

4.5.3 Cooling Source

Power conditioner chillers, absorbing chillers, or other supplies such as ambient chilling or "free conditioning" from deep lakes, borehole wells, and other resources produce chilled water at the space cooling facility. The chilling supply can be linked to the main system explicitly or implicitly via a heat exchanger(s). The direct method can only be used if fluid is the transmission medium and if the chilling supply and distribution network have the same fluid quality and pressure criteria (Meyer et al., 2013). The chilling supply and distribution network can be controlled as distinct systems with distinct temperatures and pressures, which gives both units additional flexibility in design.

For space conditioning, there are primarily two types of cooling systems. (1) absorbing coolers and (2) vapour-compression unit coolers. Vapour-compression energy, turbines, or internal combustion engines can all be used to power vapour-compression coolers.

In centrally cooling water uses, electrically powered (rotational or rotary compressor) coolers are most frequent. Pneumatic coolers would use R22, R-134a, R-123, or ammonia as refrigerants (in positive displacement machines).

4.5.4 Absorption Chillers

Lithium bromide (LiBr)–water and aqueous ammonia mixtures are the two most prevalent absorption methods used in commercially produced absorption chillers. The liquid is the refrigerant in the lithium bromide–liquid solution, whereas the ammonia is the refrigerant in the ammonia–liquid solution. The refrigeration source used determines the minimum cooling water output temperature changes that can be achieved by an absorption chiller. The LiBr chiller can create water that is as cold as 4 °C. Cooling towers relying on ammonium water can create cooling water or media at temperatures considerably below 0 °C. The number of "actions" or "phases" can be used to define the absorption chiller. The amount of heat intake into the cooler is known as the impact. The number of evaporator/absorber combinations operating in multiple temperature ranges within an absorption chiller is referred to as the number of stages.

4.5.5 Distribution System

Both the velocity of flowing liquid and the thermal gradient influence the thermodynamic efficiency of the space air conditioning system. Space cooled liquid is delivered to consumers via supply tubes from chilling source(s) and recovered after extracting heat from the house's secondary cooled water supplies. The cooled water is distributed by pumps, which create a differential pressure (DP) between the source and reverse pipes. At the crucial junction of the network, the pump head (PH) is chosen to counteract the turbulence in the source and return pipes, as well as the pressure difference in the client location or power transmission station (PC) (Bloomquist, 2003).

The volume of water that runs through every building ETS is governed by one or more check valves with a wide flow operational range responding to fluctuations in the requirement for chilling in the buildings. For high temperature demands, the feed temperature for space refrigeration towers varies from 7 to 12 °C. Under low loads, certain space chillers will accept a feed temperature of up to 10 °C.

4.5.6 Energy Transfer Station (ETS)

The ETS refers to the interaction between the space refrigeration/heating system and the house air conditioning. Separation and controlling valves, regulators, measurement equipment, an energy analyser, and heat exchangers are all part of the ETS. It is possible to create the network with both direct and indirect connections. The cooling energy water is transferred immediately to a wired network within the house, including air management and heating coils, induction modules, and so on, using a direct relationship. Here, between the space system and construction techniques, an indirect link uses one or more heat exchangers.

4.5.7 Geothermal Heating

Subsurface water pumped through the below-ground pipe system is returned to the heat exchanger (evaporator end) unit within the home during the heating process. It subsequently goes via the refrigeration field heat exchanger throughout the aquifer system. Heat is transferred to the coolant, which heats and condenses to form a low-temperature vapour. The underground water then flows out and is dumped into a lake or through a well in an open platform. The refrigerant is fed back into the subsurface pipe network to be warmed in a closed loop arrangement. The reverse valve sends liquid refrigerant to the compressor, which compresses it and reduces its volume, causing it to warm up. In the evaporation, different fluids (such as R134a and CO_2) are employed according to the operational heating rates.

4.5.8 Heat Exchangers

For air conditioning, the heat exchanger is an essential element of the thermal transit station. As a result, heat exchangers must be able to produce the amount of heat duty, temperature difference ("T"), and pressure difference (DP) according to the specifications. Heat exchangers, also called plate-and-frame heat exchangers or flat plate heat exchangers, are the only form of heat exchangers that can offer a low-temperature strategy (less than 1 °C) and are used in space air conditioning units. Another kind of heat exchanger without a gasket that could be utilized for smaller areas is the tack welded plate heat exchanger. Other types of heat exchangers, such as shell and tube or shell and coil, aren't normally suited for space chilling operations because they cannot meet the required temperature-controlled approach.

4.5.9 Control Unit

Another important part of the ETS is the main controller. To guarantee optimal control performance, special attention should be paid to the choice of check valve and control method. The controller's goal is to keep the cooled water source temperature right for the consumer while also providing a high yield to the process definition. While maintaining an appropriate indoor temperature, this control method will save power costs by optimizing settings for the check valve. For operational, surveillance, and data collection at the ETS, microcomputer electronic systems, either digital signal controller (DSC) or programmable logical control (PLC), are employed.

4.5.10 Energy Meters

The energy sensor keeps track of how much energy is transmitted from the customer's secondary unit to the main unit. The combination of mass flow, temperature differential, water temperature, and duration equals cooling energy. Because measuring mass flow in a confined tube system is challenging, volume flow is measured instead. The result is adjusted to account for the volume and heat capacity of the liquid, both of which are temperature dependent. Pressure has such a small impact that it can be overlooked. A flow sensor, a set of temperature analysers, and an energy analyser that combines the stream, temperature records, and adjustment factors make up an energy meter. The energy sensor should be delivered as a complete instrument, properly validated with specified accuracy graduation rates and in accordance with established metering regulations.

4.6 Case Studies

4.6.1 Bakreswar

Bakreswar is located in the Birbhum district of West Bengal (23°52′48″ N, 87°22′40″ E). Bakreswar is located in the eastern section of the Indian peninsula, in the far east of the fault-constrained Sone–Narmada–Tapti (SONATA) rift region, which runs from west to east across the nation (GSI, 1991; Shanker, 1988).

4.6.1.1 Methods

Bakreswar water and gas specimens were tested for chemical features and the Bakreswar region's radioactive signature. The chemical composition of water specimens was analysed using an atomic absorption spectrometer. At the Bakreswar geochemistry surveillance lab, helium and radon gases were measured using technology including a micro gas chromatograph as well as an Alpha Guard radon detector. MATLAB® 2011 was used to study the data and flatten the collected time series analysis to estimate average helium and radon concentrations and reservoir temperature. Geothermometers are used to determine the temperature of the underground fluid reservoir. A silica thermometer and an Na–K–Ca geothermometer are utilized in this experiment. The Kalina cycle-driven binary power station can be used to leverage electricity generation in Bakreswar-Tantloi. The energy provided by a heat exchanger utilizing organic working fluid in the Kalina cycle is primarily ammonia, which is erosive and corrosive in character. As a result, a Kalina-based dual power station has been suggested for the Bakreswar thermal region to reduce turbine lifespan losses. In the Kalina cycle, the organic fluid flow is an ammonia–water combination with a very low boiling point. The binary cycle technique is by far the most efficient way to generate power from low-temperature sources. Before being gathered in pre-washed polyethylene containers, the specimens were filtered on site over a 0.4-m membrane. Cation and silica (SiO_2) contents were measured using samples stored in different containers with concentrated HNO_3. A standardized thermometer was used to monitor the water's temperature change in the field. At 25 °C, a pH meter (ELICO, L1–120) was used to test pH on site, and a conductivity meter (ELICO, CM180) was used to assess conductivity. Colorimetry using a computerized spectrophotometer was used to measure SiO_2 content (ELICO, GS 5700A). Atomic absorption spectrophotometry was utilized to determine the proportions of the cations Na, K, Ca, Mg, Al, Cu, Zn, and Fe (Perkin Elmer, 306). SO_4 was measured with a turbidimetric approach using barium chloride as a solution in a Nephelo Turbidity Meter 131. For Cl measurement, the materials were determined by titration with $AgNO_3$ in the presence of a K_2CrO_4 indicator. The chemical properties of waters were plotted using a trilinear graph, which revealed that groundwaters are primarily Ca–HCO_3 kinds. A ternary Cl–SO_4–HCO_3 graph was generated for Bakreswar's groundwater sources and geothermal hot springs (Giggenbach, 1992). The graph of Cl versus tritium revealed that geothermal springs differ significantly from groundwater sources. A series of thermal springs with comparable physicochemical properties (particularly Cl concentrations) but varying issuing temperatures is expected to cool through conduction throughout ascent. Temperature versus SiO_2 also was obtained by plotting, with the points indicating that thermal springs lie between the solubility graphs of quartz and amorphous SiO_2, asserting that perhaps the evolving geothermal waters are fully saturated with quartz at surface and near-surface temperatures, and saturated at deeper temperatures. A series of hydrolysis processes and their estimated values have been obtained to examine the chemical equilibria among groundwater and soil/rock (Chandrasekhram and Chandrasekhar, 2010). SOLMINEQ 88 is a computer program that calculates the main aqueous constituents. The Giggenbach (1988) Na–K–Mg trilinear equilibria diagram shows that such fluids are not suited for determining significant Na–K and K–Mg equilibrium temperatures.

Thermal and non-thermal samples were taken from seven various sites in the Bakreswar thermal region, comprising monsoon (M), pre-monsoon (PM), and post-monsoon (POM). The specimens were taken in disinfected 1000-mL plastic bottles and delivered according to the American Public Health Association's established protocols. Millipore liquid was used to make up all quantitative reagents and references. In the field, a mercury thermometer was used to measure temperature. A computerized pH meter (Orion; Thermo Fisher Scientific, Waltham, USA) was used to detect the pH of specimens, and a conductance bridge was used to test electrical conductivity (EC) (Systronics, Norcross, USA). The amount of dissolved oxygen (DO) in the air was determined with an Orion multi-parameter analyser DO electrode. The redox potential (ORP) was determined using the redox electrode of an Orion multi-parameter analyser. The TDS (mg/L) was measured with a Systronics water analysis kit model (WA-371). The ions Ca^{2+}, Mg^{2+}, Co_3^{2-}, Na^+, HCO_3^-, K^+, Cl^-, $NH4^+$, SO_4^{2-}, F^-, and PO_4^{3-} were analysed using an ion chromatograph (IC-Plus,5, Metrohm). The Statistical Package for the Social Sciences (IBM SPSS, V.20) program was used in this work to convert the analysed data for the implementation of multivariate approaches. In addition, the geochemical indexes and graphs for the analysed data were studied using Aqua-Chem, V.2010.1, and Origin Pro, V.8 program. The saturation index (SI) of several mineral processes was determined using the PHREEQC V.3 program. ESRI's ArcGIS V.10.3 software was used to create maps of samples collected and local geology. To assess the spatial patterns of significant ions in the study area, the measurement values were merged into a geographic information system (GIS) with a Kriging geo-statistical framework.

4.6.1.2 Societal Benefits

Using available technology, the Bakreswar thermal region might be used as a continuous and consistent heat source for a long time to run a power plant with a power output of 38–76 MW.

4.6.2 Tatapani

Surguja district, Chhattisgarh, Tatapani (23°41′12″N, 83°39′04″E) is a prospective warm water reserve in central India situated in the Son-Narmada boundary. In Tatapani, a series of hot springs erupts in a 5-km-wide valley.

4.6.2.1 Methods

Water specimens were taken from Tatapani, and physio-chemical tests were performed using the Piper graph plot. Transition processes such as K_2O–Al_2O_3–SiO_2–H_2O and N_2O–Al_2O_3–SiO_2–H_2O demonstrate the disintegration of silicates. The flow charts were created at a number of different temperatures (Saxena and Gupta, 1986). The chemical structure of geothermal water from Tatapani hot springs and bore wells was investigated in the Chemical Laboratory of the Geological Survey of India (GSI) Central Region (Misissale et al., 2000). A quartz chalcedony geothermometer is used to determine the temperature of the aquifer. An elongated E-W trending telluric low was discovered during an audio magnetic telluric survey at Tatapani, suggesting the existence of a conductivity region in this area (Harinarayan, 1998). In conjunction with a water reservoir with a width of roughly 3 km, the Tatapani MT anomaly is linked to a narrow conductive fault/fracture region that extends to deeper depths (Harinarayan, 1998). In a few specimens, fluid inclusion investigations on primary and secondary biphase intrusions are observed. Hydrothermal modification can be seen predominantly around the Tatapani thrust fault. The presence of extensive silica sinter in the area indicates that the thermal liquid was of the alkali chloride kind (Sarolkar and Mukhopadhyay, 1998).

For the investigation of environmental tritium, water samples were taken from hot springs, groundwater, and rivers in 1-L polyethylene bottles. Water specimens were taken and processed via 0.45-mm pore diameter filters before being distilled at atmospheric pressure, avoiding interaction with the atmosphere. To control the electrolytic procedure, 250 mL of distilled water was placed in an electrolytic cell with 0.5 g of Na_2O_2, an electrolyte that produces NaOH. The cells are derived from the refrigeration unit after the electrolysis program is finished and measured for their capacity. $PbCl_2$ is added to the electrolysed sample water to reduce it further. The final drained and electrolytically enhanced water is placed in polythene vials with liquid scintillator cocktail (Optiphase Hisafe3) in an 8:14 ratio and recorded for 500 minutes using a Perkin Elmer 1220 Quantulus scintillation counter. National Institute of Standards and Technology (NIST) references are used to recalibrate the counter. The role in determining the outcome was 0.5 tritium units (TU) (primary criterion for analytical error). The 3H/1H ratio of 1 TU of sample is 10–18, which translates to 0.12 Bq/L of water or 3.2 pCi/kg H_2O. Temperature, EC, and pH were all evaluated in situ utilizing Hanna's multi-parameter system (HI 9828). To comprehend both shallow and deep circulating fluids, an EC versus tritium graphic was used.

4.6.2.2 Social Benefits

The Tatapani thermal region is central India's largest potential geothermal asset. National Thermal Power Corporation (NTPC) has been awarded permission by the state government to construct a geothermal power station in the Tatapani area of Balrampur district. This will be the nation's first geothermal energy plant. According to preliminary results, the cost of production for 300 kWe would be $0.14/kWh and $0.1/kwh for 5 MWe capacity, assuming a 95% production installed capacity and a rate of interest of 8% excluding tax liability.

4.6.3 Iceland Deep Drilling Project

Iceland is comprised of basalt, which was produced by volcanic activity across the Mid-Atlantic Ridge (MAR). The North Atlantic Ocean first formed 56 million years ago. Greenland and Scandinavia were gradually slipping apart. Rifting occurred throughout millions of years, and at roughly 20 Ma, the MAR and Iceland plumes approached each other, eventually covering the hot spot. Iceland's basaltic plateau was forming at this time. Greenland to the west and Eurasia to the east are receding at a rate of about 2 centimetres each year. The plumes and oceanic crust spreading centre have moved eastwards, resulting in many rift zones.

4.6.3.1 Method

The partnership intended to dig 4–5-km-deep wells from the commencement of the process, and the Iceland Deep Drilling Project (IDDP) well was constructed at a depth of 3.5 to 4.5 km (Thórhallsson et al., 2003; Pálsson et al., 2014). Pressure and temperature measurements from prior well locations at Reykjanes, Krafla, and Hengill were used to create the primary well design (Thórhallsson et al., 2014). IDDP opted to dig at Krafla in 2007 using five cemented casing strings and a7" slotted liner ultimate well diameter. For the top 300 meters of the anchoring casing, API T-95 was used in addition to the standard steel grade K55. As it has better compressive and tensile strength than a Buttress link, a so-called "Wedge Thread" or Hydril 500 type was selected for casing attachments rather than a Buttress thread (Thórhallsson et al., 2014). The wellhead must be able to endure, at approximately 470 °C, 180 bar of significantly lowering and 220 bar of shut-in pressure under projected high p-T circumstances (Pálsson et al., 2014). Except for the control valve body, which is constructed on a specific class section of 250 bar, all elements can withstand a minimum pressure of 420 bar. The blowout preventers and bits are the same as those used in traditional geothermal installations. To cover the sidewalls and generate a filter cake, bentonite drilling fluid was utilized. A mud chilling device was installed in the mud systemic circulation. When digging for anchoring casing, mud motors were added to the bottom hole assembly (BHA) to create more mechanical power (Pálsson et al., 2014).

As they reached the final level of 2103 m, the digging crew ran into a lot of issues. A coring operation using a customized core barrel proved unsuccessful. The first drilling effort resulted in a damaged BHA and a 30-meter-long part of equipment 9 consisting of the Anderdrift, two stabilizers, an 8-inch DC, and the bit (Pálsson et al., 2014). Harvesting and acidizing procedures were unsuccessful; therefore, the fish was left in the hole and filled with cement. The design has had to be changed due to unforeseen challenges. The anchoring casing was moved 400 meters up to 2000 meters, and a sidetrack was created at 1934 meters. The interject was drilled to a depth of 2096 meters, where it encountered complete loss of circulation as well as a significant increase in torque. Finally, the BHA became caught in the magma once more; it was cemented, and a new sidetrack was completed. The drilling crew chose to disengage the mud engine at that moment to reduce economic risk (Pálsson et al., 2014). The drilling procedure began again, but this time there was a complete loss of flow. The hooking load abruptly fell by 45 tons at 2096 m deep, and the pipe became forever trapped for the third time. The executive board agreed to finish the well at 1935 meters with 95/8″ production tubing and a 95/8″ slotted liner below to 2072 meters. The interior diameter of the producing liner is larger than that of a standard7″ oil and gas liner. This allows optimal power output while also lowering pressure losses at high flow rates (Bello and Teodoriu, 2012). Following that, the production casing received reverse and equal circulation work from the outside. Initially, the cement slurry was pushed down into the annulus and replenished the water. The next stage was to use trial and error to calculate a static water level.

The water was pushed up at the same volume speed as the cement slurry, and then the two were balanced. The TOC (top of the cement) was recorded at a level of 725 m, with the basement of the cement at 1700 m. TOC was 3 to 4 m below ground after a second cement operation (Pálsson et al., 2014). Circulation tests were undertaken in autumn 2009, and they discovered several issues with the master valve as well as material deterioration.

References

1. Ait Ouali, A., Ayadi, A., Maizi, D., Issaadi, A., Ouali, S., Bouzidi, K., Imessaad, K., 2020, Updating of the Most Important Algerian Geothermal Provinces, In *Proceedings, World Geothermal Congress 2020*, Reykjavik, Iceland, 8 p.

2. Al-Shammiri M, Safar M., 1999, Multi-effect Distillation Plants: State of the Art, *Desalination*, Vol. 126, pp. 45–59.
3. Andritsos, N., Dalampakis, P., Kolios, N., 2003, Use of Geothermal Energy for Tomato Drying, *GHC Bulletin*, March.
4. Arason, S., 2003, The Drying of Fish and Utilization of Geothermal Energy; the Icelandic Experience. Presented at "International Geothermal Conference", Reykjavík, Iceland, 11 pp.
5. Bakema, G., Provoost, M., Schoof, F., 2020, Netherlands Country Update, In *Proceedings, World Geothermal Congress 2020*, Reykjavik, Iceland, 11 p.
6. Bello, O., Teodoriu, C., 2012, Development of a Model to Calculate Rig Power Requirements for Geothermal Applications, Paper SPE 162958 Presented at the Nigeria International Conference and Exhibition, Abuja, 6–8 August, 2012.
7. Bloomquist, G.R., 2003, Geothermal Space Heating, *Geothermics*, Vol. 32 (4–6), pp. 513–526.
8. Buros, O.K., 2000, *The ABCs of Desalting*, 2nd edn. ASIN: B0006S2DHY, International Desalination Association, USA, p. 30.
9. Buros, O.K., 1999, *The ABCs of Desalting*, International Desalination Association, USA, p. 31.
10. Chandrasekharam, D., Chandrasekhar, V., 2010, Hot Dry Rock Potential in India: Future Road Map to Make India Energy Independent, *World Geothermal Congress 2010*, Bali, Indonesia, 25–29 April 2010.
11. Daysh, S., Carey, B., Doorman, P., Luketin, K., White, B., Zarrouk, S.J., 2020, 2015–2020 New Zealand Country Update, In *Proceedings, World Geothermal Congress 2020*, Reykjavik, Iceland, 17 p.
12. El-Nashar, A.M., 2001, Cogeneration for Power and Desalination – State of the Art Review, *Desalination*, Vol. 134, pp. 7–28.
13. Fytikas, M., Andritsos, N., Karydakis, G., Kolios, N., Mendrinos, D., Papachristou, M., 2000, Geothermal Exploration and Development Activities in Greece during 1995–1999, In *Proceedings of World Geothermal Congress 2000*, edited by S. Rybach et al., Kyushu-Tohoku, Japan, 28 May – 10 June.
14. Giggenbach, W.F., 1988, Isotopic Shift in Waters from Geothermal and Volcanic Systems along Convergent Plate Boundaries and Their Origin, *Earth and Planetary Science Letters*, Vol. 113, pp. 495–510.
15. Goldbrunner, J., 2020, Austria – Country Update, In *Proceedings, World Geothermal Congress 2020*, Reykjavik, Iceland, 19 p.
16. Gondwe, K., Mwagomba, T., Tsokonombwe, G., Eliyasi, C., Khumbolawo, L., 2020, Geothermal Development in Malawi – a Country Update 2015–2020, In *Proceedings, World Geothermal Congress 2020*, Reykjavik, Iceland, 5 p.
17. GSI, 1991, *Geothermal Atlas of India*, GSI special pub no. 19, Geological Survey of India, Nagpur, India, pp. 102–110.
18. Hamed, O., Mustafa, G.M., BaMardouf, K., 2001, Prospects of Improving Energy Consumption Of The Multi-stage Flash Distillation Process, In *Proceedings of the Fourth Annual Workshop on Water Conservation in Dhahran, the Kingdom of Saudi Arabia*, WGC.
19. Harinarayan, T., Sarma, S.V.S., 1998, Deep Electrical Conductivity Investigations in some Geothermal Areas of India, In *Deep Electromagnetic Exploration*, edited by K.K. Roy, Narosa Publishing House, New Delhi, pp. 160–175.

20. Kallin, J., 2019, Geothermal Energy Use, Country Update for Finland, In *Proceedings, European Geothermal Congress 2019*, European Geothermal Congress, Den Haag, The Netherlands, 5 p.
21. Kaushik, G., Nand, S., 2015, District Cooling Conversion System: A Case Study, *International Journal of Engineering*, Vol. 3 (4), pp. 678–683.
22. Khawaji, A.D., Kutubkhanah, I.K., Wie, J-M., 2008, Advances in Seawater Desalination Technologies, *Desalination*, Vol. 221, pp. 47–69.
23. Líndal, B., 1992, Review of Industrial Applications of Geothermal Energy and Future Considerations, *Geothermics*, Vol. 21 (5/6), pp. 591–604.
24. Lund, G., 1995, *Dairy Processing Handbook*, Tetra Pak Processing Systems AB, p. 442.
25. Lund, J.W., 2000, *Direct Heat Utilization of Geothermal Resources*, Geo-Heat Center, Oregon Institute of Technology, Klamath Falls, OR.
26. Meyer, D., Wong, C., Engel, F., Krumdieck, S., 2013, Design and Build of a 1 Kilowatt Organic Rankine Cycle Power Generator, In *Proceedings 35th New Zealand Geothermal Workshop*, WGC, New Zealand, pp. 1–7.
27. Misissale, A., Vaselli, O., Chandrasekharam, D., Magro, G., Tassi, F., Casiglia, A., 2000, Origin and Evolution of "Intracratonic" Thermal Fluids from Central–Western Peninsular India, *Earth and Planetary Science Letters*, Vol. 181, pp. 377–314.
28. Mujundar, A.S., 1987, *Handbook of Industrial Drying*, Marcel Dekker Inc., New York.
29. Pálsson, B., Hólmgeirsson, S., Guðmundsson, Á., Bóasson, H.Á., Inagson, K., Sverrisson, H., Thórhallsson, S., 2014, Drilling of the Well IDDP-1, *Geothermics*, Vol. 49, pp. 23–30.
30. Perko, B., 2011, Effect of Prolonged Storage on Microbiological Quality of Raw Milk, *Mljekarstvo*, Vol. 2 (61), pp. 114–124.
31. Phetteplace, G., 2007, Geothermal Heat Pumps, *Journal of Energy Engineering*, Vol. 133, pp. 32–38.
32. Raluy, R.G., Serra, L., Uche, J., 2004, Life-cycle Assessment of Desalination Technologies Integrated with Energy Production Systems, *Desalination*, Vol. 167, pp. 445–458.
33. Sarolkar, P.B., Mukhopadhyay, D.K., 1998, Geochemistry of Thermal Water from Production Wells, Tatapani Geothermal Field, India, *GRC Transactions*, Vol. 22, pp. 129–135.
34. Saxena, V.K., Gupta, M.L.,1986, Geochemistry of the Thermal Waters of Salbardi and Tatapani, India, *Geothermics*, Vol. 15, pp. 5–6.
35. Shanker, R., 1988, Heat-flow of India and Discussion on Its Geological and Economic Significance, *Indian Minerals*, Vol. 42, pp. 89–110.
36. Shi, J., Le Maguer, M., Kakuda, Y., Liptay, A., Niekamp, F., 1999, Lycopene Degradation and Isomerization in Tomato Dehydration, *Food Research International*, Vol. 32, pp. 15–21.
37. Thórhallsson, S., Matthíasson, M., Gíslason, T., Ingason, K.,Pálsson, B., Fridleifsson, G.Ó., 2003, Part II: Drilling Technology, In Iceland Deep Drilling Project, Feasibility Report, edited by G.Ó. Fridleifsson. Available at: www.iddp.is/wp-content/uploads/2003/11/Feasibility-Report/IDDPFR-Part-2.pdf (10 June 2014).

38. Thórhallsson, S., Pálsson, B., Hólmgeirsson, S., Inagson, K., Matthíasson, M., Bóasson, H.Á., Sverrisson, H., 2014, Well Design for the Iceland Deep Drilling Project (IDDP), *Geothermics*, Vol. 49, pp. 16–22.
39. Tian, T., Zhang, W., Wei, J., Jin, H., 2020, Rapid Development of China's Geothermal Industry – China National Report of the 2020 World Geothermal Conference, In *Proceedings, World Geothermal Congress, 2020*, Reykjavik, Iceland, 9 p.
40. Torkar, K.G., Golc Teger, S., 2008, The Microbiological Quality of Raw Milk after Introducing the Two Days' Milk Collection System, *Acta Agriculturae Slovenica*, Vol. 1 (92), pp. 61–74.
41. Wangnick, K., 2000, IDA Worldwide Desalting Plants Inventory: Report No. 16. Produced by Wangnick Consulting for the International Desalination Association.
42. *Water Desalination Technologies in the ESCWA Member Countries*, 2001, UN-ESSWA, ASIN: E/ESCWA/TECH/2001/3. United Nations, New York, p. 172.
43. Winter, T., Pannell, D., McCann, L., 2005, The Economics of Desalination and Its Potential Application in Australia, internal report, University of Western Australia.
44. Yasukawa, K., Nishikawa, N., Sasada, M., Okumura, T., 2020, Country Update of Japan, In *Proceedings, World Geothermal Congress 2020*, Reykjavik, Iceland, 7 p.

5

Machine Learning Techniques for Geothermal Energy Sector

5.1 Introduction to Machine Learning Techniques

In almost every sector, there has been an explosion of data over the last few decades. This amount of data will be meaningless unless an analytical tool is utilized to organize and infer the trends inherent within the databases. ML is one such way to process and analyse immense datasets. ML is a method for converting information and data into knowledge (Diersen et al., 2011). Based on a definition by Mitchell (1997), ML is technically "a computer program to learn from experience 'E' with respect to a desired set of tasks 'T' and performance measure 'P', if its performance at tasks in 'T', as measured by 'P', improves with experience 'E'." For an accurate classification approach, ML processes can be grouped into supervised methods, unsupervised methods, or reinforcement methods. Although every one of these procedures has its own series of requirements and tactics, the basic concept is the same. For instance, the active learning frequently includes data collection and processing, in which information is gathered and organized into a suitable state before going on to the most important stage, training. The ML algorithm is trained on the information source at this point. After that, a modelling assessment is done to determine how well the system performed on a specific dataset. Subsequently, tuning is done to improve the performance of the algorithm (Geron, 2017). The procedure for this technique is depicted in Figure 5.1.

Artificial neural networks (ANNs) are among the most promising ML techniques, which can be used to process vast quantities of data and extract useful insights. An ANN is a collection of artificially produced neurons that are taught to accomplish objectives using only a set of techniques. The basic structure of a biological neuron is adapted into the central principle of ANNs. The ANN is composed of three phases. The input layer allows data units to flow into the network, which are then passed onto the hidden layer; the convolution layer of an ANN collects input units and separates the much more relevant information bits depending on parameters; the output unit is the final step of a neural net, and it provides the network's

DOI: 10.1201/9781003204671-5

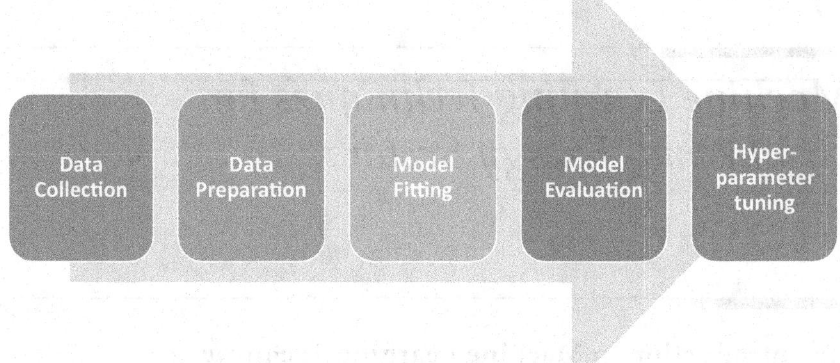

FIGURE 5.1
Steps involved in machine learning (ML) for data analysis.

forecasts. Each of the layers described here is concatenated together. One of the most crucial components of a prudent neural network is data. Significance, volume, structure, and disturbance are all features of a database that have a direct effect on an ANN's positive outcome. Significance, volume, structure, and disturbance are all features of a database that have a direct effect on an ANN's positive outcome. Another important feature of an ANN is learning, as proper training helps the channel's accuracy (Shahin, 2016).

A neuron basically transforms input into an individual output, which can then be transmitted to subsequent neurons. Figure 5.2 shows the fundamental concept of a neuron, in which each piece of a profile is amplified by its average measure and then combined to produce a total output. Each neuron uses the learning algorithm to communicate the data into a net result. Neurons in a feedforward neural network (FNN) are arrayed in such a way that they really do not interact with neurons in the very same level; rather, they are linked to cells in the neighbouring layer, as seen in Figure 5.2. Beginning at the activation function, the flow of data is unilateral in this arrangement. Weight values are generally obtained by finding the variance among outcomes after the inputs have already been processed through the network and the associated output signal patterns have been found.

The goal of an FFN is to simulate the linear function f*. In a predictor, for example, y = f*(x) transfers an input x to a class y. An FNN finds the values of the variables that result in superior faster convergence by defining a map y = f(x;) and learning the set of parameters that produce a positive outcomes optimization algorithm. Back-propagation is a technique for

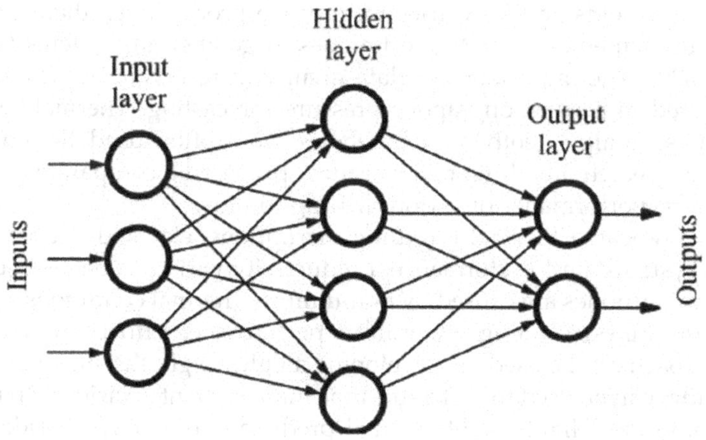

FIGURE 5.2
Fundamentals of neural network.

adjusting weight. The method uses the preceding epoch's error margin (i.e. iteration). We can lower error needs and increase the model's generalization by fine-tuning the parameters, rendering it more precise. Back-propagation is the abbreviation for "backward propagation of mistakes." It's a popular method of training ANNs. This method is useful for determining the gradients of a weight vector when all the channel's weights are taken into account. The "backwards" component of the term alludes to the idea of measuring the gradient backwards from across networks, starting with the gradient from the last layer of values and ending with the gradient of the very first layer of values.

5.2 Application of Machine Learning Technique in Estimation of Geothermal Temperature Distribution

ML is also being used for geothermal energy and geosciences. ML encompasses a diverse collection of approaches that are applicable to a variety of domains in various fields. The growth of ML technology for geothermal energy is projected to open up major prospects for integrating emerging technologies and lowering costs over the project life cycle, including geothermal exploration and power plant operations. Advanced exploration research for geological, geochemical, geophysical, and geothermal analysis using ML methods offers useful insights about geothermal energy. Despite the fact that ML techniques have been commonly adopted by different fields and industries to improve productivity, there exists a lack of implementation of ML techniques in geothermal systems (Yadav et al., 2020). The importance of data quality in reservoir ML research is highlighted in a work on vapour pressure forecasting. Thermal streams and high-enthalpy geothermal boreholes were often used to construct valuable research. In addition, the authors prepared a comparative evaluation of the performance of various ML algorithms.

Due to repeated thermal throttling disruptions, elevated temperatures restrict systems' best performance. Productivity variances between essentially similar nodes are caused by non-uniform thermal conditions around virtual machines, resulting in significant performance reduction overall. An ML approach can be used for developing a lightweight thermal prediction model that can be used to make run-time management decisions. To investigate optimized lightweight thermal predictors, one may consider two techniques. To begin, employ feature selection techniques to optimize the effectiveness of previously developed ML techniques. Furthermore, to perform a complete assessment of prediction systems, alternative methodologies based on deep neural networks and regression analysis may be developed. Ramos and Bianchini (2008) proposed a model for estimating the temperature consequences of different temperature control policies. Because their model relies on heat exchange equations, a full understanding of the underlying infrastructure architecture and its thermomechanical characteristics is required. Implementation sequencing on a product results in a distinct thermal profile, according to Yang et al. (2008). They developed a kernel-level temperature planner and proposed a strategy for anticipating such behaviour. Their heuristic-based model is designed to reduce dynamic thermal control within core components. It doesn't account for high-performance computing (HPC) operations. To evaluate thermal characteristics and develop a network heat flux for data centres, one can configure a simulation that considers node layout, power utilization, and coolant setup. The design can be tailored to specific system parameters, such as cooling type, power usage measurements, and nodal type; however, it takes a very long time to create. Some researchers have presented a model in which they merely use the information provided to the nodes and do not require any special data directly.

Moore et al. (2006) described a method for developing a data centre thermal analysis using data from exterior thermometers, server equipment, and server rooms' air conditioners. The system employs a neural network-dependent technique. This approach, which has no understanding of real implementations, is eventually utilized to assign a power budget via the data centre. Therefore, the loads in the algorithm have little learning component. The new framework efficiency for forecasting is utilized in order of the linear function, specified in Eq. (5.1), to represent the temperature properties of the system. It's important to note that no detailed knowledge of the basic process is assumed.

$$P_j(i) = f_j(A(i), A(i-1), P(i-1)) \tag{5.1}$$

Where P_j is the temperature property of the system and $A(i)$ is the associated enthalpy.

For instance, the system is unconcerned about node location and systems design. Moreover, no property heat modelling information is used in the process. This indicates that the structure is unaware of the heat transfer qualities of the components. It functions as a simple linear model between operating system (OS)-accessible functions and the expected heat at a location. Moreover, some of the approaches, such as ML and Bayesian networks, showed instability according to researchers. Linear regression algorithms functioned well, according to the researchers, especially for smaller forecast windows. Until the forecast time frame surpasses 25 seconds, the Gaussian process offers the highest predictive performance of all the techniques.

Regression analysis, Gaussian process regression, and a multilayer neural network system were all investigated by the authors. These techniques are more useful for determining whether the methodologies are suitable for a dynamic predictive model. An expansion in the time window was accompanied by an increase in mistakes. In addition, when compared with the earlier outcomes, the ML algorithm was accurate in this study. Furthermore, the three techniques performed quickly in terms of mistakes at the 20-second mark. The Gaussian mixture system is the finest model when the forecast time exceeds 30 seconds, while the proposed algorithm was a close second.

5.3 Subsurface Characterization Using Machine Learning Technique

Due to the advancement of ML in the field of computer science, various studies have attempted to implement ML models to the oil and gas industry to help enhance subsurface characterization, produce pseudo petrophysical data, and characterize crack-bearing materials. Numerous studies have demonstrated that using ML models to enhance subsurface characterization will yield positive results (Li, 2019). ML algorithms have been used to perform tasks such as well log analysis and synthetic log generation. The digitization method in the oil and gas industry facilitates the implementation of ML models. ML algorithms were used in numerous studies to build synthetic logs. Typically, these tasks entail a basic regression problem. Because of its simplicity, the ANN model is a common choice for such tasks.

Geothermal researchers utilized downhole temperature readings from big oil and gas field databases to create thermal heat flow and temperature-at-depth models in terms of finding potentially geothermally hot locations. XGBoost provides the best reliability for subterranean temperature prediction, with mean average and root-mean square-error of 3.19(°C) and 4.94(°C), correspondingly.

Furthermore, utilizing the XGBoost model, this method is constructed to incorporate two-dimensional (2D) constant temperature graphs at three distinct altitudes, which can be used to highlight potential geothermal energy active areas around the locations. Deep neural network (DNN), Ridge regression (R-reg), and decision-tree-based (e.g. XGBoost and random forest) models are the primarily used regression models in such approaches. Ensemble-based methods (such as random forest and XGBoost) were primarily used, since they are assumed to perform well in situations where there is a class imbalance (Galar et al., 2011). Furthermore, tree-based algorithms usually increase accuracy by lowering prediction variance.

Several researchers developed a method for predicting potential values of well characteristics based on the training of neural networks. TensorFlow is one of the best frameworks to apply the chosen ML methods to identify the relationships between temperature time series and historical reservoir data, future well flow rates, and future injection temperatures. MLP, LSTM, and CNN were three separate neural network architectures that were used and tested by the authors. These networks were first put to the test on a single-fracture geothermal system that had a single injection well and one production well, for which an analytical solution can be found. In preliminary analysis, it was discovered that integrating physical parameters that would be accessible to geothermal operators, such as injection temperature and mass flow, resulted in more accurate predictions instead of using a pure time series forecasting method. As a result, based on past production temperatures, injection temperatures, and well injection mass flow rate, future production temperatures are predicted. The authors used the signal from two additional channels (Figure 5.3) to amplify the temperature training sequences of length N = 12 (tuneable parameter) in the single-fracture modelling problem: Injection temperature and injection mass flow (both variables are regulated by the geothermal reservoir operators).

An ML technique, such as Kernel Ridge Regression (KRR), can be built and trained using synthetic and/or existing seismic data. In cognitive

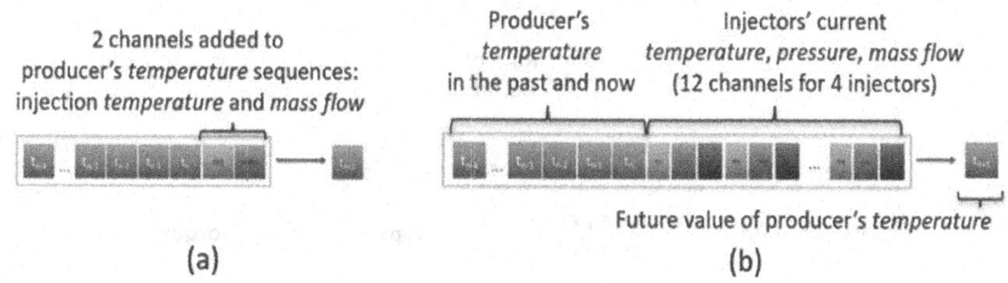

FIGURE 5.3
Machine learning algorithm for time series prediction developed for a simple single-fracture doublet system (a) and for a BHS model with four active injector wells and six active production wells (b).

training, a function, f*, is created that takes a d-dimensional measurements array as input and generates a label forecast.

For any unknown measurement vector $x' \in R^d$, we can predict its label f*(x'). This method is used to characterize subsurface geological features. The training algorithms are used to generate a mapping function, f(x'), which establishes a relationship between the seismic data and the geological features.

The accuracy of this ML approach, KRR, can be justified by ML theory (Mohri et al. 2012). The prediction error reduces monotonically when larger datasets are used for training.

5.4 Application of Machine Learning Technique in Geothermal Drilling

One of the most critical factors in a well's performance is its rate of penetration (ROP). This is evidenced by the increasing number of studies on the subject. In numerous research articles, however, the methodology is disputed (Hegde et al., 2017). For instance, simple models correlating ROP and RPM have been proposed in numerous studies (Maurer, 1962 (ref. 19); Bourgoyne et al., 1986). A good ROP model for roller cone parts can be formulated by considering the various factors that affect the bit size, operating conditions, and rock strength. For instance, if the bit size is small, then the bit ROP is proportional to the bit size (Hareland and Rampersad,1993; Jahanbakhshi, 2012).

Many analytical relationships on ROP estimation have been proposed over time, and since the breakthroughs in the field of data science and ML, scientists have implemented many data-driven models, including ML

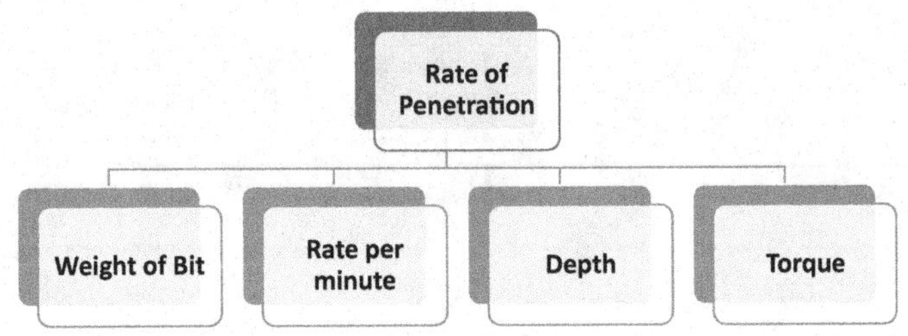

FIGURE 5.4
Machine learning scheme for rate of penetration.

techniques, that have shown suitable results in addition to higher accuracy (Hegde et al., 2017).

The set of input parameters and the input data are the only factors that influence ML predictions. Therefore, after the models have been adequately trained with input datasets, they can be expected to be usable when drilling by calculating surface parameters for real-time ROP estimation. ANN, deep learning, random forests, and other regression and classification approaches are among the ML techniques that have numerous applications in drilling and have been shown to be effective.

Numerous researchers have studied the drilling rate of forests using ML methods (Breiman, 1996). Some of these include random forests (Bilgesu et al., 1997; Bataee et al., 2011; Hegde et al., 2017), decision trees (Buntine, 1992), and support vector machines (Ahmed et al., 2018).

Many ROP models have been developed to predict the rate of penetration in wells (Figure 5.4). ML techniques can help in estimating this parameter. A study conducted in Iran revealed that ML could accurately predict the penetration rate in a well (Rashdi et al., 2020). The problem is presented by measuring the ROP prediction using four input parameters: the weight of bit, the revolutions per minute, the drilling depth, and the torque.

A tree method can be used to identify drilling conditions in an area. However, analytical ROP models can only segment the response of a single factor (Hegde et al., 2015). The conventional trees are not very accurate in predicting ROP in a field drilling scenario (Soares and Gray, 2019). Deeper trees have a high variance and can significantly alter the tree structure. An algorithm may expand several deep trees with varying data points and then average out the estimates for each tree for each data point (Soares and Gray, 2019).

Ahmed et al. (2018) used support vector machines (SVM) to predict the ROP in a shale formation. They collected 400 points in the formation to test their model. The optimal dispersion for the information was identified to be 70% for learning and 30% for assessment. The networks were validated by using a learning algorithm but instead evaluated using the evaluation metrics, which had never been utilized previously. The SVM model was trained and tested using the following input variables: load on bit, standpipe stress, string rotational velocity, stream pump, thrust, mud density, funnel stiffness, elastic modulus, yield point, and particulates (%).

5.5 Application of Machine Learning Technique in Reservoir Modelling

One of the most challenging subjects of geothermal reservoir engineering is reservoir classification and prediction modelling. Since cold wastewater injection can cause thermal breakthrough in producers, we need to consider how injection affects production in order to maintain our geothermal fields sustainably. The role of predicting thermal breakthrough is critical in the management of a sustainable geothermal reservoir. Characterizing the reservoir and creating a geological model are key applications in this prediction task. Although identifying fractures and flow paths in geothermal reservoirs is difficult, it provides useful information for effective geothermal reservoir development. Though challenging, flow paths in geothermal reservoirs provide useful knowledge for efficient geothermal reservoir growth. Understanding the layout of fractures and their connectivity helps determine optimum injection strategies for reservoirs because fractures basically control mass and heat transport in the system. Prediction modelling traditionally entails creating a computational model of the reservoir based on discovery data and characterization efforts, and then transforming that model into a reservoir simulation model, which is then calibrated using inverse analysis and production data. Prediction modelling in the conventional sense is often a time-consuming and computationally monotonous operation.

Several studies on ML and deep learning specifically for modelling subsurface behaviour, for example, have recently been published. In a single-well pressure transient study, Liu and Horne (2013) suggested a kernel-based convolution model to predict a pressure response. In their research, input features were carefully designed using flow rate and time transformations. This concept was then further expanded by Tian and Horne (2015) to incorporate pressure analysis for multi-well systems. Kernel-ridge regression was used instead of kernel-based convolution and was able to minimize computing costs while increasing prediction accuracy. In addition, deep learning techniques like recurrent neural networks could classify the reservoir model and map the relationship between pressure and flow rates without modifying the input features (Tian and Horne, 2017).

Li et al. (2019) used temperature and flow rate data as inputs to predict downhole pressure, demonstrating the strengths of recurrent neural networks. As opposed to flow rates and choke valve opening, temperature was the best parameter for pressure prediction in their research. Gudmundsdottir and Horne (2020) examine how neural networks can be used to model the interaction between injectors and producers in a geothermal reservoir without relying on a physical model. A model like this is advantageous when it comes to making quick predictions for reservoir management; if accurate, it enables one to study and run various well control scenarios to optimize for the reservoir's most sustainable exploitation. A feedforward neural network and a recurrent neural network are two deep learning approaches compared by the authors. For the contrast, a synthetic geothermal reservoir was numerically simulated, and data from four injectors and five producers were used as input and output for the neural networks. The emphasis was on two cases in particular: Creating different models for each producer and creating a single model that predicts all producers at once. The basic feedforward neural network known as Multilayer Perceptron (MLP) outperformed the more flexible structure of the recurrent neural network in this simulated geothermal reservoir. The addition of the target feed improved the models' efficiency substantially, and the authors were able to simplify the model design in the case of MLP by decreasing the number of hidden layers. The authors observed that in comparison to the recurrent neural network (RNN), the MLP was therefore much faster to learn because there were fewer parameters to predict. In this study, an MLP had 10,901 parameters to prepare for a neural network of 50 neurons and 5 hidden layers, whereas an RNN had 23,001 parameters. When compared with the training of the MLP, the RNN takes five times longer to train.

Mohamed et al. (2020) (Reservoir 3) suggested a deterministic artificial neural inversion approach for forecasting reservoir attributes. Several ML techniques were verified and proved to accurately model various reservoir features. The probabilistic neural network (PNN) is another of those techniques. The PNN could be used to forecast reservoir parameters as well as classify them. Employing data from external boreholes and seismic data, the approach trains an algorithm to forecast reservoir features. The three major types in the procedure are identifying the proper collection of characteristics that use the non-linear regression approach. The researchers use non-linear regression during the first phase to find the number of characteristics that effectively predicts the targeted reservoir characteristic, that is, the set of characteristics that properly predicts a given basin characteristic with the least posterior probability. The researchers use prestack inversion dimensions as improved yield in this stage, in addition to the seismic obtained (internal) characteristics. Seismic obtained properties include immediate, windowed bandwidth, filtering slice, derivation, averaged, and temporal characteristics. The researchers put every one of these characteristics to the evaluation and then apply statistical modelling to determine the best order and number of characteristics that yield the lowest model complexity, which is then preserved for the training process. Within the second phase, the PNN technique is designed to train the classifier. Gaussian weighting functions are used to apply the seismic properties to the learning basin characteristics data. By sprinting each mean of such curves at every training set, the method optimizes the standard error or smoothing variable, sigma, for each characteristic. Some sigma levels are determined automatically by the cross-validation procedure. The researchers erase a third of the borehole data from the training phase mostly during the verification test, which occurs concurrently with the learning, and help to determine these boreholes randomly. Finally, utilizing three-dimensional (3D) seismic and inverted dimensions, the training set is used to estimate the targeted resource attribute.

A seismic full stack pre-stack height relocation area recorded at 4 ms with a recorded length of 6 s could be used as information in this research. The in-house pre-stack seismic inversion dimensions P-impedance, S-impedance, P-wave velocity, S-wave acceleration, VP /VS proportion, lambda-rho, and mu-rho could also be employed. The interpretative data focuses on the structural images (channel top, gas bottom, and stream base), fault markers, and fault triangles. Lowered modelling approaches can be utilized to minimize the simulation duration of extremely large subsurface flow simulations. All the processes employ appropriate orthogonal deconstruction, which requires conducting an increased training simulation, saving solution pictures, and applying an eigen-decomposition to the generated given dataset. During oil reservoir simulation, formulas regulating the movement of reservoir fluids (oil, gas, and water) via a permeable geological formation must be solved. In broader versions, individual topics (e.g. methane, ethane, etc.) and/or thermal impacts are recorded. Oil–water flows, for example, are analysed in the absence of a catalyst. Further enhanced productivity could be contemplated with modest alterations if the read-only memory (ROM) is used in a simulation that is very general. The boundary condition for oil, gas, and water movement in porous media is determined using Darcy's law and mass transfer (Aziz and Settari, 1986). The entirely implicit result of the reservoir simulation study is Eq. (5.2):

$$T^{n+1}X^{n+1} - D\left(X^{n+1} - X^{n}\right) - Q = R \tag{5.2}$$

T represents a block Penta diagonal matrix for 2D grids and a block heptadiagonal matrix for 3D grids, D represents a block diagonal matrix, Q represents the source/sink terms, and R represents the residual vector. The superscript n or $n + 1$ designates the time standard. Convective effects are represented by the $T^{n+1} X^{n+1}$ term, while accumulation is represented by the $D(x^{n+1} - x^n)$ term. The matrices T, D, and Q are all dependent on x and must be modified at every time stage iteration.

Equation (5.2) is non-linear and is solved by using Newton's equation, $J\delta = -R$, to drive the residuals to zero, where J is a Jacobian matrix given by $J_{ij} = \partial R_i / \partial x_j$

And $\delta_i = x_i^{n+1,k+1} - x_i^{n+1,k}$, where k and $k + 1$ represent iteration level.

5.6 Application of Machine Learning Technique in Geothermal Power Generation

5.6.1 Prediction of Geographical Suitability of Geothermal Power Plants

Geothermal energy is increasingly being used as a source of electricity for both household and commercial purposes in a growing number of nations. Geothermal power plants assist in reducing global warming by generating electricity from a natural, renewable resource in a sustainable manner. The viability of a location for geothermal energy generation, which is a dynamic and unpredictable mix of various environmental conditions, determines the efficiency of a power plant. Geothermal suitability assessments now require invasive inspections, significant expenses, and legal approvals. Therefore, having a global suitability map of geothermal plants as a reference during inspections would be useful prior knowledge, saving time and resources. A worldwide suitability map will enhance understanding of the link between future geothermal resource usage and geothermal life cycle sustainability in the Life Cycle Assessment (LCA) approach (Tomasini-Montenegro et al., 2017; Parisi et al., 2019; Karlsdottir et al., 2020). Numerous studies have been conducted in the past that combined data from a wide range of sources to give open-access overviews of geothermal infrastructure at the local or country level (Moghaddama et al., 2014; Trumpy et al., 2015; Noorollahi et al., 2007). In recent decades, regional suitability maps for various types of renewable energy resources have been produced by mapping natural resources onto operating plants using geographic information system (GIS) software (Angelis-Dimakis et al., 2011; Mourmouris and Potolias, 2013; Bertermann et al., 2014). In addition, various analytical approaches have assessed the appropriateness, efficiency, and continuous availability of renewable energy in specific countries or areas by applying data envelope analysis and decision trees (Ramazankhani et al., 2016; Martín-Gamboa et al., 2015; Karimi et al., 2019; Troldborg et al., 2014).

These studies also consider socio-economic factors such as population density, industries, access to skilled labour, and so on, as well as environmental issues that are very specific to the region being studied and are not often available for other regions, even though they would be useful in these cases. As a result, such methods are unlikely to be relevant to other regional or global assessments. They also use models that factor parameters (by making judgments about one or a few parameters at a time) or that use linear parameter combinations. Predicting ideal geothermal locations, on the other hand, would need a more sophisticated technique: Scalable decision-making systems with a wide range of parameters. Many researchers have utilized ML approaches (such as ANNs, SVM, random forests, and others) to predict the link between specific characteristics and the potential of a region for geothermal resource exploitation (Santoyo et al., 2018; Shahab and Singh, 2019). These operations, too, rely on a few factors that are particular to the region under study.

Maximum Entropy (MaxEnt) is an ML approach that is commonly used in ecological modelling to predict species distributions (Baldwin, 2009; Coro et al., 2015; Coro et al., 2018). In several circumstances, MaxEnt may be used to estimate a probability density function px using real-valued vectors. The x vectors might be environmental factors related to the presence of a species or the frequency of a phenomenon (Pearson et al., 2012; Coro et al., 2018 (ref. 53)). This approach is sophisticated enough to exclusively deal with positive examples (presence-only model). Because the vectors only relate to the locations where the phenomenon happens in this case, non-occurrence is determined automatically. The model may be trained using current geothermal power plants as positive examples, with geothermal site suitability as the phenomenon to model. The model can utilize the environmental factors associated with the regions where geothermal plants operate as input vectors. The MaxEnt implementation, which is offered as a service on the D4Science e-Infrastructure, may be used to train the model (CNR, 2019; Phillips et al., 2019). The model's parameters are modified by the training algorithm if the density function is consistent with pre-set mean values at the training-set locations and the density function's entropy $H = \sum \pi(\overline{x})\ln((\pi(\overline{x})))$ is highest there (Elith et al., 2011). As training sites, locations of recognized geothermal power plants were chosen. With respect to the entropy values of the parameters of random points selected all over the world, MaxEnt conducts a relative maximization of the entropy function on these places. The methodology uses a linear combination of parameters, with coefficients modified to reflect the relative relevance of each factor in predicting training-set locations (percentage contribution). The output of the model is also influenced by the permutation of variable values recorded by the training algorithm inside the training vectors (permutation importance). After the training phase, percentage contribution may be used to choose the most important environmental features, that is, MaxEnt can be used as a filter to select the most informative parameters. Factors with a contribution value more than 5% of the maximum contribution value, for example, may be maintained to improve model performance or to create a new projection based exclusively on the most significant parameters (Coro et al., 2013). A MaxEnt model trained on 0.5 resolution parameters may be used to produce probability estimates for the environmental parameters of 0.5 squared regions. Based on the data in the training set, probability might be considered as a suitability factor for geothermal power plants. The disadvantage of MaxEnt is that it is particularly susceptible to dataset bias; thus, its performance increases if the training locations are reliable. To improve the model's robustness, only real data from geothermal power plants should be utilized for training (Coro and Trumpy, 2020).

5.7 Challenges and Issues in Adapting Machine Learning for the Geothermal Energy Sector

As the amount of data has exploded in recent decades, several industries, including the geothermal sector, are seeking to capitalize on this new resource in order to improve their efficiency. ML is utilized in almost every field, from manufacturing to regulatory activities to scientific, health, and security applications (Chen, 2012). Furthermore, many businesses use ML techniques integrated in their information systems to improve the quality of their operations or to offer new services and products (Schüritz et al., 2016; Dinges et al., 2015). The broad application of ML, on the other hand, is still in its early phases and hence faces several challenges. High-dimensional datasets (Cai et al., 2018), algorithm scalability (Hazelwood et al., 2018), distributed computation (Zhou, 2017), and live ML on streaming data are all important issues that have lately surfaced in research and practice (Zhou, 2017). According to certain research, when developed algorithms are applied in real-world implementations, output disparities diminish swiftly (Rudin et al., 2014). Furthermore, although analysis performs evaluations on real-world databases, it is suggested that the findings are not relayed to the application domain in a timely way (Wagstaff, 2012).

Data is the starting point for each ML project. A good data structure with the right consistency and enough data samples are required for a successful project. When only a small amount of data is available, difficulties may occur. Even when data is accessible, obtaining satisfactory results is always challenging due to missing data, incorrect entries, or noisy features (Kocheturov et al., 2018). Another issue in the ML domain is unbalanced or biased data. Even though a variety of solutions have been proposed, there is still room for improvement ((Blenk et al., 2017; Zhou et al., 2016). ML on encrypted data sets is equally tough. Despite significant progress on issues like training on small, encrypted datasets, more work must be done when it comes to training on large, encrypted datasets).

Big data, on the other hand, has its own set of issues. In certain aspects, using training algorithms on massive amounts of data might be difficult (Saidulu et al., 2017). It's especially challenging to deal with big volumes of data in a timely manner (Lopes et al., 2017).

Furthermore, the large dimensionality of datasets in big data systems demands more complicated feature engineering and other pre-processing methods (Sarwate et al., 2013; Ferguson, 2018). Transfer learning for applying models to related activities across domains can also be problematic due to differing data distributions (Suthaharan, 2014). In this case, balancing previously observed patterns against new training data is tough. The concept of technical debt, which refers to the extra effort necessary in the future to modify unclean code compared with clean code (Sculley et al., 2015), is another issue that arises throughout the modelling phase. This is an issue that arises during the pre-deployment phase, but it can also affect the deployment phase.

Infrastructure issues, such as reducing model training computation effort and therefore lowering memory and energy consumption, as well as increasing training and efficiency levels, are commonly addressed concerns in model design (Saidulu et al., 2017). Large data volumes aggravate these problems. Solutions must be explored to make the models useable (Dietterich et al., 2008). Furthermore, when employing ML models, data privacy and security are governance problems that must be considered (Lopes et al., 2017). Furthermore, the increased dimensionality of datasets in big data systems demands more sophisticated feature engineering as well as extra pre-processing techniques. Additionally, transfer learning for applying models to similar activities across domains might be challenging due to differing data distributions. It's tough to weight previously acquired patterns with new training data in this situation.

Another common hindrance for ML applications in sectors like geothermal energy is gathering appropriate data. This is a significant issue because the availability, quality, and composition (e.g. is the meta-data included? Is the dataset labelled?) of the acquired dataset available has a strong influence on the performance of ML algorithms. Currently, the majority of ML methods only operate with data that contains continuous and nominal values. The method and parameter settings, among other factors, determine the magnitude of the impact. The inability to access specific data is a challenge for most industrial research, not only ML applications, due to concerns such as security or a lack of data gathering during the process. Even while ML allows the extraction of information and delivers better results than most traditional techniques with fewer data needs, certain features of the available data must be considered to enable a successful implementation.

5.8 Case Studies

5.8.1 Case Study 1: Drilling

The authors (Rashdi et al., 2020) propose Ensemble Bagged Trees, an ML algorithm for predicting the ROP in formations using data from the weight on bit (WOB), rotary speed (RPM), torque, and measured depth. A big well segment in Iran was included in this analysis, and there was no information break in the segment. ROP can be conceived of as a function of numerous variables, including surface parameters, rock strength, geology, lithology, porosity, bit design and form, drilling mud, human factors, downhole conditions, mud rheology, and other associated parameters. A portion of a well was drilled in Iran, and the data collected during this drilling was collected by the authors and used as the training data for ROP prediction. Though analytical ROP models can't segment a single variable's response, decision trees can recognize drilling conditions in operational regions given by multiple drilling features (Soares and Gray, 2019). According to the authors, simple trees would likely perform poorly in predicting ROP in a practical field drilling scenario. More internal nodes that partition the dataset into additional regions should be added. Deeper trees, on the other hand, have a high variance, which means that a slight alteration in the training data will drastically alter the tree structure and result in significantly different predictions. The researchers developed algorithms that expand several deep trees based on a subset of the data (with bootstrap sampling) and then average out the estimates of all trees for each data point to address high variance.

The following are the findings of the Ensemble of Bagged Trees algorithm, which the authors used for their study with a minimum leaf size of 8 and a total of 30 learners. On testing results, a substantial level of accuracy was achieved, with a correlation coefficient (R^2) of 0.8 and a mean squared error (MSE) of 0.803.

Both values had good accuracy as compared with previously used ROP prediction models, which is a significant advantage over most of the previously introduced ROP models. This study demonstrates how Bagged

Decision Trees (Random Forests Extension) can accurately predict the values of ROP using surface measured parameters as input parameters.

5.8.2 Case Study 2: Fault Detection

In a range of organizations and disciplines, ML methods have been employed to develop fault diagnosis systems. ML techniques are used by Zulkarnain et al. (2019) to build a fault diagnosis classifier model for geothermal power plants with essential motors. The researchers' approach is a comparison of techniques that provide the highest accuracy indicator of classes using a simple classifier and an ambient learner.

In order to create a more reliable model, data is pre-processed before being used to build a fault detection model. Aggregating data, removing missed values, deleting outliers (sensor errors), and scaling data are also examples of data pre-processing. Scaling data means normalizing all variables to 0–1 with the help of a min–max normalization formula. Dataset scalability is often used to minimize weighting fluctuations in classifiers or groupings that are susceptible to size variations, which would make the system inaccurate. After pre-processing, the information is reduced to 3963 cases, which will be utilized to develop the system. The information assets used in this research were based on methods used to assess the true situation of the liquid plant. This value comes in several different data scales and units. In four different methods, types of numeric values are used as independent factors or influencing forces. To indicate the status, cluster analysis forms might be employed. The factors are listed in Table 5.1.

When developing a classifier, it's critical to determine the condition of each occurrence. The requirement is determined using the K-Means grouping algorithm. The goal of data grouping is to appropriately group data with regular conditions and alarms. To commence creating a classifier model, the organizations will be labelled for each occurrence. The Euclidean distance with such a K value of 2 is used in K-Means grouping. The information is categorized based on the results of the K-Means group till the categorization model is developed. The grouping method was

TABLE 5.1

Factors Required for Fault Detection System

Attributes	Units
Electrical currents	Ampere (Amp)
Electrical resistance	Ampere (Amp)
Pressure	Barg (Bg)
Heat	Degrees Celsius (°C)

assessed using the Davies–Bouldin (DB) indices. The DB index is a variable used to validate the cluster model that was created. The index ranges from 0 to 1, with a value near 0 signifying a more effective clustering approach. This data was used to predict the density of datasets in a group as well as the range among cluster centres. The density of datasets in a group and the length among cluster centres were measured using this approach.

ANNs are quite well known for their self-learning capabilities in dynamic interconnections, which are inherent in fault classification and diagnosing difficulties, and also their simplicity of use. The Back-Propagation Neural Network (BPNN) is an inter-convolutional neural network learning technique. SVMs are guided learning algorithms for performing regression and classification evaluation. Researchers have developed SVM gradient descent (SGD) in their investigations. SGD is renowned for making SVM both strong and easy to configure. SGD is affected by multiple scales of characteristics. SVM solution in fault identification provides generalized effectiveness even with low training data. SVM is a technique for detecting and diagnosing problems in components like bearings and industrial machinery. The group (boosting) software XGBoost is a gradient boosting solution and utility. The enhancing strategy necessitates building the forest in a stepwise order, with each succeeding tree seeking to minimize the variance of the preceding tree. XGBoost is a new ML method with a huge range of design variables that beats existing algorithms on the basis of availability and reliability.

Researchers classify the findings into two groups: first, classifier indications based on data learning, and second, classifier indicators based on data assessment. This contrast is intended to show which algorithm produces an underfitting prediction. The defect feature representation was also validated using the monitoring and research findings. It's worth noting that the very same median should be used when labelling data for processing as for grouping data for training with the K-Means clustering methods.

References

1. Ahmed, S., Elkatatny, A., Abdulrahim, M., Mahmoud, A., Ali., 2018. Prediction of rate of Penetration of Deep and Tight Formation Using Support Vector Machine. Society of Petroleum Engineers. SPE-192316-MS.
2. Angelis-Dimakis, A., Biberacher, M., Dominguez, J., Fiorese, G., Gadocha, S., Gnansounou, E., Guariso, G., Kartalidis, A., Panichelli, L., Pinedo, I.. Robba, M., 2011, Methods and Tools to Evaluate the Availability of Renewable Energy Sources, *Renewable and Sustainable Energy Reviews*, Vol. 15, pp. 1182–1200.

3. Aziz, K., Settari, A., 1986, *Fundamentals of Reservoir Simulation,* Elsevier Applied Science Publishers, UK.
4. Baldwin, R.A., 2009, Use of Maximum Entropy Modelling in Wildlife Research, *Entropy,* pp. 854–866. Available at: https://doi.org/10.3390/e11040854.
5. Bataee, M., Mohseni, S., 2011, Application of Artificial Intelligence System in ROP Optimization: a Case Study in Shadegan Oil Field. Society of Petroleum Engineering. SPE 140029.
6. Bertermann, D., Klug, H., Morper-Busch, L., 2015, A Pan-European Planning Basis for Estimating the Very Shallow Geothermal Energy Potentials. *Renewable Energy,* Vol. 75, pp. 335–347.
7. Bilgesu, H., Tetric, L.T., Altmis, U., Mohaghegh, S., Ameri, S., 1997, A New Approach for the Prediction of Rate of Penetration (ROP) Values. Society of Petroleum Engineering. SPE 39231.
8. Blenk, P., Kalmbach, S., Schmid., 2017, o'zapft is: Tap Your Network Algorithm's Big Data!. Conference: ACM SIGCOMM 2017 Workshop on Big Data Analytics and Machine Learning for Data Communication Networks (Big-DAMA), Los Angeles, CA, USA.
9. Bourgoyne, K., Millhiem, M., Chenevert, F., Young., F.S., 1986, *Applied Drilling Engineering,* SPE Textbook Series, Society of Petroleum Engineers.
10. Brieman, L., 1996, Bagging Predictors. Department of Statistics, University of California. Technical Report No. 421.
11. Buntine, W., 1992, Learning classification trees, Statistics and Computing, Vol. 2, pp. 63–73.
12. Cai, J., Luo, J., Wang, S., Yang, S., 2018, Feature Selection in Machine Learning: A New Perspective, *Neurocomputing,* Vol. 300, pp. 70–79.
13. Chen, H., Chiang, R.H.L., Storey, V.C., 2012, Business Intelligence and Analytics: From Big Data to Big Impact. Management Information Systems Research Center, University of Minnesota, Minnesota. Available at (stable URL): www.jstor.org/stable/41703503.
14. CNR, 2019. Maximum Entropy Model Web Processing Service. https://services.d4science.org/group/biodiversitylab/data-miner?OperatorId¼org.gcube.dataanalysis.wps.statisticalmanager.synchserver.mappedclasses.transducerers. MAX_ENT_NICHE_MODELLING.
15. Coro, G., Trumpy, E., 2020, Predicting Geographical Suitability of Geothermal Power Plants, *Proceedings of Journal of Cleaner Production,* Vol. 267, pp. 1–9.
16. Coro, G., Magliozzi, C., Ellenbroek, A., Pagano, P., 2015, Improving Data Quality to Build a Robust Distribution Model for *Architeuthis dux, Ecological Modelling,* Vol. 305, pp. 29–39.
17. Coro, G., Pagano, P., Ellenbroek, A., 2013, Combining Simulated Expert Knowledge with Neural Networks to Produce Ecological Niche Models for *Latimeria chalumnae, Ecological Modelling,* Vol. 268, pp. 55–63.
18. Coro, G., Vilas, L., Magliozzi, C., Ellenbroek, A., Scarponi, P., Pagano, P., 2018, Forecasting the Ongoing Invasion of *Lagocephalus sceleratus* in the Mediterranean Sea, *Ecological Modelling,* Vol. 371, pp. 37–49.
19. Diersen, S., Lee, E.J., Spears, D., Chen, P., Wang, L., 2011, Classification of Seismic Windows Using Artificial Neural Networks, In *Proceedings of the International Conference on Computational Science, ICCS 2011,* Elsevier.

20. Dietterich, T.G., Domingos, P., Getoor, L., Muggleton, S., Tadepalli, P., 2008, Structured Machine Learning: The Next Ten Years, *Machine Learning*, Vol. 73, pp. 3–23. Available at: https://link.springer.com/content/pdf/10.1007/s10994-008-5079-1.pdf.
21. Dinges, V., Urmetzer, F., Martinez, V., Zaki, M., Neely, A., 2015, *The Future of Servitization: Technologies That Will Make a Difference*, Cambridge Service Alliance, University of Cambridge, Cambridge, UK.
22. Elith, J., Phillips, S.J., Hastie, T., Dudik, M., En Chee, Y., Yates, C.J., 2011, A Statistical Explanation of MaxEnt for Ecologists, *Diversity and Distributions*, Vol. 17, pp. 43–57.
23. Ferguson, L., 2018, Machine Learning and Data Science in Soft Materials Engineering, *Journal of Physics: Condensed Matter*, Vol. 30 (4).
24. Galar, M., Frenandez, A., Barrenechea, E., Bustince, H., Herrea, F., 2011, A Review on Ensembles for Class Imbalance Problem: Bagging-, Boosting-, and Hybrid Base Approaches, *IEEE Transactions on Systems, Man, and Cybernetics, Part C (Applications and Reviews)*, Vol. 42 (4), pp. 463–484.
25. Geron, S., 2017, Hands-on Machine Learning with scikit-learn, keras, and TensorFlow: Concepts, Tools, and Techniques to Build Intelligent Systems. Available at: https://doi. org/10.1088/1361-6420/aa9581.
26. Gudmundsdottir, H., Horne, R., 2020, Prediction Modeling for Geothermal Reservoirs Using Deep Learning. In *Proceedings, 45th Workshop on Geothermal Reservoir Engineering*, Stanford University, Stanford, California, 10–12 February, 2020 SGP-TR-216, Stanford University Publication, USA.
27. Hareland, G., Rampersad, P.R., 1993, Drag – Bit Model Including Wear. Society of Petroleum Engineers. SPE 26957. pp. 657–667.
28. Hazelwood, K., Bird, S., Brooks, D., Chintala, S. Diril, U., Dzhulgakov, D., Fawzy, M., Jia, B., Jia, Y., Kalro, A., Law, J., Lee, K., Lu, J., Noordhuis, P., Smelyanskiy, M., Xiong, L., Wang, X., 2018, Applied Machine Learning at Facebook: A Datacenter Infrastructure Perspective, In *IEEE International Symposium on High Performance Computer Architecture*, pp. 620–629.
29. Hegde, C., Daigle, H., Millwater, H., Gray, K., 2017. Analysis of Rate of Penetration (ROP) Prediction in Drilling Using Physics-based and Data-driven Models, *Journal of Petroleum Science and Engineering*. PII: S0920–4105, Vol. 17, pp. 30725-8.
30. Hegde, C., Wallace, S., Gray, K., 2015, *Using Trees, Bagging, and Random Forests to Predict Rate of Penetration During Drilling*. Society of Petroleum Engineering. SPEE-176792-MS, SPE, Calgary.
31. Jahanbakhshi, R.K.R., 2012, Real-time Prediction of Rate of Penetration during Drilling Operation in Oil and Gas Wells, *Am. Rock Mech. Assoc.* 53, 127.
32. Karimi, H., Soffianian, A., Seifi, S., Pourmanafi, S., Ramin, H.., 2019, Evaluating Optimal Sites for Combined-cycle Power Plants using GIS: Comparison of Two Aggregation Methods in Iran. *International Journal of Sustainable Energy* Vol. 39 (2), pp. 101–112.
33. Karlsdottir M., Heinonen J., Palsson H., Palsson O., 2020, Life Cycle Assessment of a Geothermal Combined Heat and Power Plant Based on High Temperature Utilization, *Geothermics*, Vol. 4.

34. Kocheturov, A., Pardalos, P.M., Karakitsiou, A., 2018, Massive Datasets and Machine Learning for Computational Biomedicine: Trends and Challenges. Springer. Available at: www.proquest.com/openview/1c05d58cc1352a014affbf3f751ef544/1?pq-origsite=gscholar&cbl=25585.
35. Li, Y., Sun, R., Horne, R.., 2019, *Deep Learning for Well Data History Analysis*, Society of Petroleum Engineering. SPE-196011-MS.
36. Liu, Y., Horne, R.., 2013, *Interpreting Pressure and Flow Rate Data from Permanent Downhole Gauges Using Convolutions-kernel-based Data Mining Approaches*, Society of Petroleum Engineering. SPE 165346.
37. Lopes, N., Ribeiro, B., 2017, Novel Trends in Scaling Up Machine Learning Algorithms, In *Proceedings of 16th IEEE International Conference on Machine Learning and Applications*, pp. 632–636.
38. Martín-Gamboa, M., Iribarren, D., Dufour, J., 2015, On the Environmental Suitability of High- and Low-enthalpy Geothermal Systems, *Geothermics*, Vol. 53, pp. 27–37.
39. Maurer, W.C., 1962, The "Perfect-Cleaning" Theory of Rotary Drilling, *Journal of Petroleum Technology*, Vol. 14 (11), pp. 1270–1274.
40. Mitchell, T.M., 1997, *Machine Learning*, McGraw-Hill, New York.
41. Moghaddama, M., Samadzadegana, F., Noorollahi, Y., Sharifia, M., Itoi, R., 2014, Spatial Analysis and Multi-criteria Decision Making for Regional-scale Geothermal Favorability Map, *Geothermics*, pp. 189–201.
42. Mohamed, A.I., Othman, A., Fathy, M.,2020. A New Approach to Improve Reservoir Modelling via Machine Learning. *The Leading Edge*, Vol. 39, pp. 170–175.
43. Mohri, M., Rostamizadeh, A., Talwalkar, A., 2012, *Foundations of Machine Learning*, Massachusetts Institute of Technology, OK.
44. Moore, J., Chase, J., Ranganathan, P., 2006, Weatherman: Automated, Online and Predictive Thermal Mapping and Management for Data Centers, In *IEEE International Conference on Autonomic Computing*, June.
45. Mourmouris, J.C., Potolias, C., 2013, A Multi-criteria Methodology for Energy Planning and Developing Renewable Energy Sources at a Regional Level: A Case Study Thassos, Greece. *Energy Policy*, Vol. 52, pp. 522–530.
46. Noorollahi, Y., Itoi, R., Fujii, H., Toshiaki, T., 2007, GIS Model for Geothermal Resource Exploration in Akita and Iwate Prefectures, Northern Japan, *Computers & Geosciences*, Vol. 33 (8), pp. 1008–1021.
47. Parisi, M., Ferrara, N., Torsello, L., Basosi, R., 2019, Life Cycle Assessment of Atmospheric Emission Profiles of the Italian Geothermal Power Plants, *Journal of Cleaner Production*, Vol. 234, pp. 881–894.
48. Pearson, R.G., 2012, Species' Distribution Modeling for Conservation Educators and Practitioners. American Museum of Natural History, Center For Biodiversity and Conservation. Available at (stable URL): ncep.amnh.org/linc/.
49. Phillips, S.J., Miroslav, D., E, S.R., 2019, Maxent Software for Modeling Species Niches and Distributions (Version 3.4.1). Available at: http://biodiversityinformatics.amnh.org/open_source/maxent/.
50. Ramazankhani, M., Mostafaeipour, A., Hosseininasab, H., Fakhrzad, M., 2016, Feasibility of Geothermal Power Assisted Hydrogen Production in Iran, *International Journal of Hydrogen Energy*, Vol. 41, pp. 18351–18369.

51. Ramos, L., Bianchini, R., 2008, C-Oracle: Predictive Thermal Management for Data Centers, In *Symposium on High-Performance Computer Architecture.*
52. Rashdi, M., Asadi, A., Abbasi, A., Asadi, E., 2020, Machine Learning's Application in Estimation of the Drilling Rate of Penetration – A Case Study from a Wellbore in Iran, In *Proceedings of ARMA-CUPB Geothermal International Conference.*
53. Rudin, C., Wagstaff, K.L., 2014. Machine Learning for Science and Society, *Machine Language,* Vol. 95 (1), pp. 1–9. Available at: https://link.springer.com/article/10.1007/s10994-013-5425-9.
54. Saidulu, D., Sasikala, R., 2017, Machine Learning and Statistical Approaches for Big Data: Issues, Challenges and Research Directions. *International Journal of Applied Engineering Research,* Vol. 12, pp. 11691–11699.
55. Santoyo, E., Acevedo-Anicasio, A., Pérez-Zarate, D., Guevara, M., 2018, Evaluation of Artificial Neural Networks and Eddy Covariance Measurements for Modelling the CO_2 Flux Dynamics in the Acoculco Geothermal Caldera (Mexico), *International Journal of Environmental Science and Development,* Vol. 9 (10), pp. 298–302.
56. Sarwate, D., Chaudhuri, K., 2013, Signal Processing and Machine Learning with Differential Privacy, *IEEE Signal Processing Magazine,* pp. 86–94.
57. Schüritz, R., Satzger, G., 2016, Patterns of Data-Infused Business Model Innovation, In *Proceedings of IEEE 18th Conference on Business Informatics,* pp. 133–142.
58. Sculley, D., Holt, G., Golovin, D., Davydov, E., Phillips, T., Ebner, D., Chaudhary, V., Young, M., Crespo, J., Dennison, D., 2015, Hidden Technical Debt in Machine Learning Systems, *Advances in Neural Information Processing Systems,* Vol. 28, pp. 2503–2511.
59. Shahab, A., Singh, M.P., 2019, Comparative Analysis of Different Machine Learning Algorithms in Classification of Suitability of Renewable Energy Resource, In *International Conference on Communication and Signal Processing,* Patna, India, pp. 0360–0364.
60. Shahin, M.A., 2016, State-of-the-art Review of Some Artificial Intelligence Applications in Pile Foundations, *Geoscience Frontiers,* Vol. 7, pp. 33–44.
61. Soares, C., Gray, K.., 2019, Real-time Predictive Capabilities of Analytical and Machine Learning Rate of Penetration (ROP) Models, *Journal of Petroleum Science and Engineering,* Vol. 172, pp. 934–959.
62. Suthaharan, S., 2014, Big Data Classification: Problems and Challenges in Network Intrusion Prediction with Machine Learning, *Performance Evaluation Review,* Vol. 41 (4), pp. 70–73.
63. Tian, C., Horne, R.N., 2015, Applying Machine Learning Techniques to Interpret Flow Rate, Pressure and Temperature Data from Permanent Downhole Gauges. Presented at the SPE Western Regional Meeting, Garden Grove, California, USA, 27-30 April. SPE-174034-MS, pp. 386–401.
64. Tian, C., Horne, R., 2017, *Recurrent Neural Networks for Permanent Downhole Gauge Data Analysis,* Society of Petroleum Engineers. SPE-187181-MS.
65. Tomasini-Montenegro, C., Santoyo-Castelazo, E., Gujba, H., Romero, R.J., Santoyo, E., 2017, Life Cycle Assessment of Geothermal Power Generation Technologies: An Updated Review, *Applied Thermal Engineering,* Vol. 114, pp. 1119–1136.

66. Troldborg, M., Heslop, S., Hough, R., 2014, Assessing the Sustainability of Renewable Energy Technologies Using Multi-criteria Analysis: Suitability of Approach for National-scale Assessments and Associated Uncertainties, *Renewable and Sustainable Energy Reviews*, Vol. 39, pp. 1173–1184.
67. Trumpy, E., Bertani, R., Manzella, A., Sander, M., 2015, The Web-oriented Framework of the World Geothermal Production Database: A Business Intelligence Platform for Wide Data Distribution and Analysis, *Renewable Energy*, Vol. 74, pp. 379–389.
68. Wagstaff, K.L., 2012, Machine Learning That Matters, In *Proceedings of the 29th International Conference on Machine Learning*, Edinburgh, Scotland, UK, 2012.
69. Weinmann, M., Schneider, C., Brocke, J.V., 2016, Digital Nudging, Business & Information Systems Engineering, Vol. 58 (6), pp. 433–436.
70. Yadav, K., Yadav, A., Sircar, A., 2020, Feedforward Neural Network for Joint Inversion of Geophysical Data to Identify Geothermal Sweet Spots in Gandhar, Gujarat, India, *Energy Geoscience*, Vol. 2, pp. 189–200.
71. Yang, J., Zhou, X., Chrobak, M., Zhang, Y., Jin, L., 2008, Dynamic Thermal Management through Task Scheduling, In *Performance Analysis of Systems and Software*. ISPASS 2008. IEEE International Symposium on. IEEE, 2008, pp. 191–201.
72. Zhou, J., Khawaja, M.A., Li, Z., Sun, J., Wang, Y., Chen, F., 2016, Making Machine Learning Useable by Revealing Internal States Update – A Transparent Approach, *International Journal of Computational Science and Engineering*, Vol. 13, pp. 378–389.
73. Zhou, Z., 2017, Machine Learning Challenges and Impact: An Interview with Thomas Dietterich, *National Science Review*, Vol. 5 (1), pp. 54–58.
74. Zulkarnain, I., Surjandari, R., Bramasta, E., Laoh., 2019. Fault Detection System Using Machine Learning on Geothermal Power Plant. *Proceedings of IEEE*. Available at: https://ieeexplore.ieee.org/document/8887710.

6

Financing in Geothermal Energy Exploration and Exploitation

6.1 Introduction

The development and monetizing of geothermal energy in a country depend on three different parameters: Information on geothermal resources in the country, supportive policies, and robust financial models for the project developers.

Finance can have four broad elements:

a. Commercial debt/equity
b. Government grants or loan guarantees
c. Development lending
d. Financial support to combat climate change

Reliance solely on commercial capital for geothermal development is rarely viable even in the developed country market. In developing countries, where geothermal energy is a resource and is getting converted to reserve, the participation of private players is rare. In such countries the commitment of the public sector, including the country government, international donors, and financial institutions, is an essential element of success in mobilizing capital. The respective roles of the public and private sectors in financing geothermal projects depends on the desired level of vertical integration of the geothermal development market, among other factors. If private sector financing of geothermal projects is envisaged, the costs of capital need to be carefully considered, as the financiers may require a high premium for the risks involved. This is true for debt and equity funding. Debt funding typically covers the greater part of the capital requirements, up to 70% of the total capital cost: Private equity investors, however, are likely to require a relatively high rate of return on their invested capital. A required return on equity of 20–30% per year is not unusual due to the risks involved in the project.

DOI: 10.1201/9781003204671-6

In India, geothermal energy is in the nascent stage. The following stages of financing are proposed to develop geothermal projects:

a. Early-stage financing
b. Late development financing
c. Construction and operation financing

6.1.1 Early-stage Financing

The early-stage financing is for the activities shown in Figure 6.1.

This stage is the highest-risk stage and is the most difficult for raising capital. This phase has strong similarities with oil and gas exploration (Salmon et al., 2011). Three approaches that need to be followed are:

- Private equity placements of portfolios of projects
- Exchange traded corporate equity financing
- Balance sheet financing

In private equity placement financing of a portfolio of projects, geothermal companies in India need to successfully raise private equity by offering a portfolio of projects. Given the similarities in development risks to oil and gas exploration and production, private equity funds with experience in this arena may be more comfortable than others in taking on the risks of geothermal development.

India should replicate the model of North American geothermal projects. These projects are funded by exchange traded corporate equity. Projects have been financed by companies with expertise in geothermal power that filed for their initial public offering (IPC). Companies seeking to list on the public exchanges have three important competitive advantages that may enable them to raise equity at the corporate level:

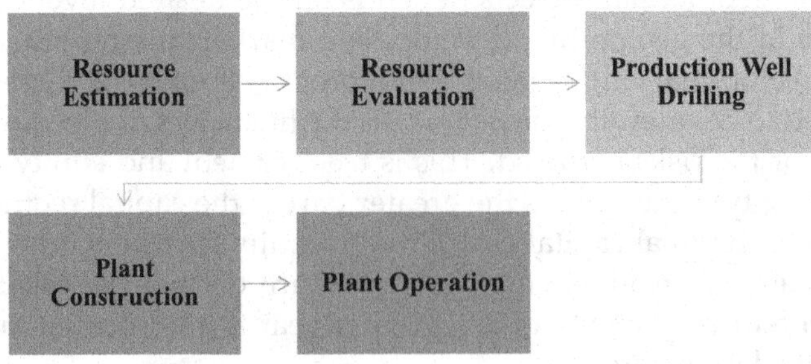

FIGURE 6.1
Early-stage financing.

TABLE 6.1

Early Stage Equity Financing Requirements

Sources	Financial Matrices	Non-financial Requirements
Public Exchanges and Private Equity	• 2× to 5× multiple on investment • Reasonable financial plan including well costing, $4–$10 million USD, success wells 2 out of 10	• Experienced management team • Ability to raise capital and carry out capital plan • Risk Management strategies
Corporate balance sheet (established market investors)	• ROE minimum 10% and preferably 13%	

- Technology expertise and experience improve the likelihood of success of any individual project.
- Project portfolios reduce the risk of an individual project by spreading the risk across an investment portfolio.
- Expectation of success in the project portfolio, which may be justified by projects located in fields or by receiving favourable reports from respected geothermal firms.

The last concept is known as balance sheet financing. Established corporations with strong balance sheets and a diverse asset portfolio that includes geothermal power plants have very different return expectations for new geothermal projects. According to market participants, while they are looking at managing higher returns from geothermal projects than from wind, they prefer a return on equity (ROE) greater than 13%, but their minimum ROE would be as low as 10%.

Table 6.1 describes the initial-stage financing requirements.

6.1.2 Late Development Financing

This stage is financed by mezzanine financing. Mezzanine financing is debt financing that includes an equity kicker. The equity upside of a project is the gain to the project developers. The leader does not have recourse to other assets of the companies. Mezzanine financing is normally provided 12 to 18 months before the construction phase. Mezzanine financing is highly risky and the financer also requires a good knowledge of geothermal project development.

6.1.3 Construction and Operating Financing

This stage of the project is shown as a flow chart in Figure 6.2.

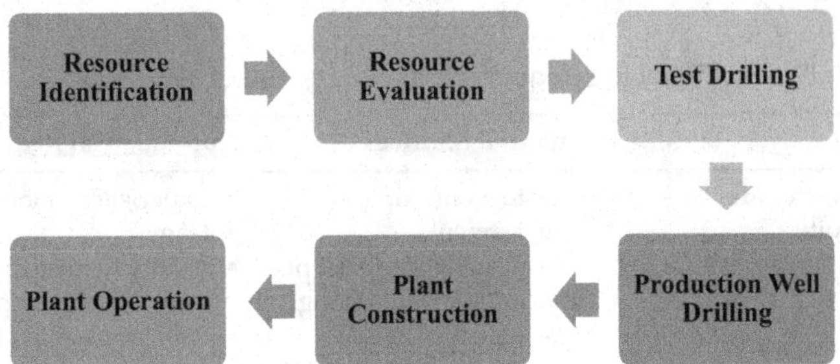

FIGURE 6.2
Financing in construction and operation stage.

This phase essentially translates into an up-front equity investment of greater magnitude. In some cases, however, it is necessary to bring in an additional equity investor as well as lenders. In the construction phase, we may have a construction loan and a term loan. The debt to equity ratio may be 55%:45% or less. A term loan guarantee requires at least 20% equity (Laenen et al., 2019).

6.2 Sources of Funding for Geothermal Projects in India

Based on a combination of secondary research, the following sources are listed for geothermal projects in India:

- Pension funds
- Tax free funds bonds
- Inflation linked bond
- Capital gains tax bonds
- Tradable tax credit
- Government/intermediary guarantees
- Infrastructure debt fund
- Renewable energy development fund

The purchase of power from geothermal energy sources, beyond the Renewable Purchase Obligations (RPO 5) stipulated by the respective state's Electricity Commission, is closely linked to the cost of supply of geothermal power and the achievement of grid parity by renewable energy sources.

6.2.1 Business and Accounting Model

Engineers and geoscientists preparing feasibility studies for geothermal projects need to include some form of financial model of the proposed project (Farhad et al., 2020). The concepts used are described here:

The monetary concept: Accountants are kept for business entities. The answer to any accounting issue must address the question: How does it affect the business? Accounting to the entity concept, accountants are not concerned with persons who own or operate the business, but with the business itself.

The entity concept: Accounting assumes that a geothermal business will continue forever. If there is evidence to the contrary, and the entity is to be liquidated, the accounting function might need to assess what the entity is worth to a potential buyer.

The dual aspect concept: The areas owned by a business or entity are called geothermal assets. The claims against this asset are called equities. The two types of equities are liabilities that represent claims of creditors and the owner's equity. The total claims on the assets are liabilities liability and stakeholder's equity.

Accounting period concept: Accounting practice is based on the need to report, periodically, the status of the business entity. The basic time period is the fiscal year (12 months). Many companies use interim (usually quarterly and/or monthly) reports.

Materiality concept: Insignificant terms do not require attention. To a small entity, a $5 million asset may be a very important item, but for a big company, it may not be sufficiently material to deserve a separate balance sheet entry or even a footnote.

Conservatism concept: The principle dictates that, given the chance, an asset will be recorded at the lowest or most conservative value. As far as the Income Statement is concerned, this principle provides that potential losses be accounted for, and yet potential gains or profits are not registered unless they are realized.

Consistency concept: The consistency concept stipulates that once an entity embarks on an accounting methodology, it must be consistent in its treatment of accounting issues unless it has good reason to do otherwise.

Realization concept: The realization principle dictates that revenue should be recognized only at the time a transaction is completed with a third party or when the value is reasonably certain.

6.3 Economic Drivers of a Geothermal Project

6.3.1 Equity Funding

This is a way of raising money for company activities by selling common or preferred stock to individual or institutional investors. Private equity is an asset class consisting of equity securities and debt in operating companies that are not publicly traded on the stock exchange. There are two sources of equity funding:

- Domestic sources
- External sources

The domestic sources are domestic investors (independently or in collaboration with international investors), public utilities, dedicated government funds, and other institutional investors. The external sources include foreign investors (independently or in collaboration with domestic investors), equipment suppliers, dedicated infrastructure funds, other international equity investors, and multilateral agencies.

6.3.2 Debt Funding

This is an investment pool such as a mutual fund or exchange traded fund. A debt fund may invest in short-term funds, securitized products, money market investments, or floating rate debts. The main investing objectives of a debt fund will usually be preservation of capital and generation of income. As in an equity fund, there are two sources of debt funding:

- Domestic sources
- External sources

The key drivers for debt funding are:

- Credit strength and experience of the geothermal project or project sponsors
- Status of support system
- Governance, structure, information reporting, transparency of company
- Creation and enforceability of security
- Experience and technology dominance of the executing geothermal company

TABLE 6.2

Domestic and External Financing Sources for Geothermal Energy

Domestic Sources	External Sources
Equity	Equity
Domestic investors (independently or in collaboration with international investors)	Foreign investors
Dedicated government funding	Equipment suppliers
Other institutional investors	Dedicated infrastructure funds
	Other international equity investors
	Multilateral agencies
Debt	Debt
Domestic commercial bond (3–5 years)	International commercial bonds (7–10 years)
Domestic term lending institutions (7–10 years)	Export credit agencies (7–10 years)
Domestic bond markets (7–10-year tenor)	International bond markets (10–30 years)
Specialized infrastructure	Multilateral agencies (over 20-year tenor)

- Marketing strength and credit strength of the giver and taker
- Hedging policy
- Reliable distribution arrangement

The financing sources for domestic and external financing are shown in Table 6.2.

6.4 Basic Structure of a Financial Model

The model is based on several interconnected worksheets in a workbook (Farhad et al., 2020). The overall structure and interconnected relationships are shown in Figure 6.3.

Some of the attributes of the model are as follows (Kalimbia, 2019):

- As far as possible, input parameters would be restricted to one sheet.
- Various sections of the model would be structured with different time scales – either monthly or financial years.
- An exercise in Excel reading of expressions, proofing, and modifications of the model.
- Colour is used quite extensively to improve the readability of both the working spreadsheets and the printed output.

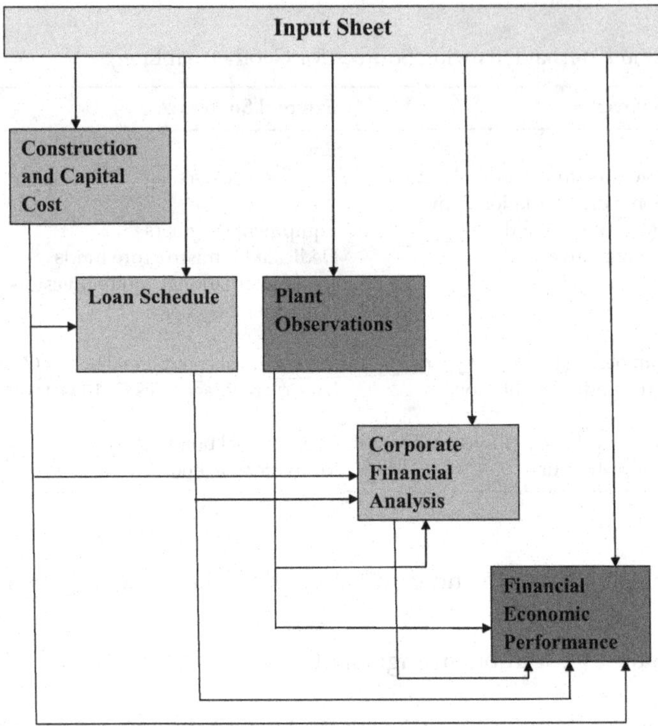

FIGURE 6.3
Basic structure of financial model.

- The model is fully documented, so that other users would be able to understand the underlying principles and calculation methodologies –and again, colour will assist in understanding the nested Excel functions.

6.5 Financial Funding in Various Countries

6.5.1 Indonesia

The World Bank's board of executive directors has approved a $150 m loan for Indonesia to scale up investments in geothermal energy by reducing the risk of early-stage exploration. The loan is accompanied by $127.5 million in grants from the Green climate funds, two institutes supporting climate friendly development. Indonesia's geothermal sector has vast potential and current installed geothermal capacity is already the second largest in the world (Michale et al., 2014).

Under the Indonesia Geothermal Resource Risk Mitigation (GREM) project, the financing will help public and private sector developers to mitigate risks in exploration of geothermal resources, including covering a part of the cost in the case of unsuccessful exploration. The project will also finance technical assistance and capacity building of key stakeholders in the geothermal sector.

To achieve the government's target of 23% renewable energy mix by 2025 requires a contribution from geothermal development of about 7% or equal to 7000 MW. It is an ambitious, huge development with a total investment of $35 billon.

Indonesia now has 1.9 GW of installed geothermal power and plans to develop an additional 4.6 GW to help meet the government's target for renewable energy.

6.5.2 The Philippines

The Philippines is a geothermal hotspot. Developing the natural resource will meet the country's growing demand for power with low carbon and fuel imports. The Tiwi and Makban geothermal Green Bonds projects are the seventh and fourth largest geothermal facilities in the world. The Asian Development Bank (ADB) lent $37.7 million to help finance the plants' operation, maintenance, and expansion. In the Philippines' first green bond issue, ADB guaranteed 75% of a $225 million equivalent bond issued. The Philippines model of investment in 2020 is shown in Figure 6.4.

6.5.3 Iceland

The Nordic Investment Bank (NIB) is the international financial institution of the Nordic and Baltic countries with headquarters in Helsinki, Finland. NIB signed a 15 years' loan with Icelandic utility Orkuveita Reykjavikur to finance an investment programme to improve geothermal power production and electricity distribution networks in Iceland. The USD 80 million loans will help finance new make-up and reinjection wells, new steam pipeline connections, and a new workshop building as well as an expansion of the thermal station in Hellsheid geothermal power plant. The driving of new make-up wells is a continuous activity to maintain the capacity of geothermal power plants. In the Nesjavelur geothermal power plant, the projects include new make-up wells, a complete control system update, and replacement of equipment in the thermal stations. The loan will also co-finance investments in the electricity distribution system, strengthening and replacing old equipment in the network, harbour electrification, and

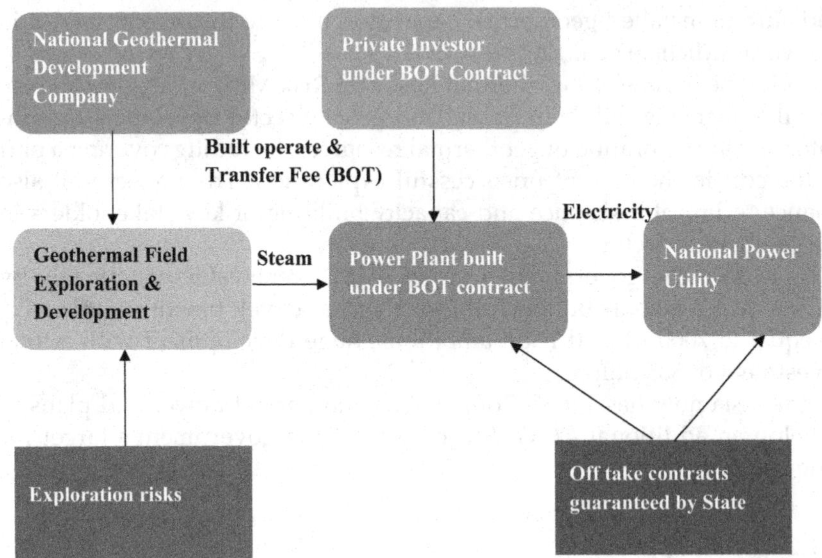

FIGURE 6.4
Investment model of the Philippines.

the installation of smart meters. The electrification of the harbour area will allow shore-to-ship electricity and thereby reduce emissions from ships running on auxiliary diesel power when moved. In 2006, NIB provided a loan to Orkuveita Reykjavikur to finance the construction of a power plant to provide electricity, district heating, cold sewage, and a fibre-optic cable network to customers in southwest Iceland. In the key segments of power and heating it serves over half of Iceland's population. The group owns and operates three power plants, the Nesjavellir and Hellisheidi geothermal plants and the Andakilsa Wirkjun hydroelectric station, with a combined capacity of 431 MW of electricity. Producers are assured a value for the electricity they deliver into utility systems when feed-in tariffs are already in effect for direct usage (space heating) and/or generated electricity from geothermal energy. According to Rybach (2010), comparable tariffs exist in numerous European nations, notably Belgium, Austria, Greece, the Czech Republic, France, Spain, Estonia, Slovenia, and Slovakia, and the policy has resulted in huge geothermal development in Germany. According to Gassner (2010), the German Renewable Energy Sources Act of 2009 requires the power supply national grid to accept and prioritize electricity generated by renewable energy as well as to pay regular prices set by law for a 20-year term. The additional costs are passed on to consumers. In this way, Gassner notes, the state itself is not involved in financing but instead merely controls the framework conditions, which allow project developers, investors, and operators to reliably calculate yields for the first 20 years of operation.

References

1. Farhad, T.H., Mortha, A., Hadi, F.A., et al., 2020, Role of Energy Finance in Geothermal Power Development in Japan, *International Review of Economics and Finance*, Vol. 70, pp. 398–412.
2. Gassner, H., 2010, Feed-in Tariffs, Support Policy and Legal Framework for Geothermal Energy in Germany, In *Proceedings World Geothermal Congress 2010*, Bali, Indonesia, 4 pp. Available at: www.geothermal-energy.org/pdf/IGAstandard/WGC/2010/0309.pdf.
3. Kalimbia, C., 2019, Financial Modelling and Analysis of Power Project Finance: A Case Study of Ngozi Geothermal Power Project, Southwest Tanzania, Report of United Nations University, pp. 1–113. Available at: https://orkustofnun.is/gogn/unu-gtp-report/UNU-GTP-2019-01.pdf.
4. Laenen, B., Bogi, A., Manzella, A., 2019, Financing Deep Geothermal Demonstration Projects, European Technology and Innovation platform on deep geothermal, pp. 1–32. Available at: www.etip-dg.eu/front/wp-content/uploads/4.5-Financing-deep-geothermal-Final-1.pdf.
5. Micale, V., Oliver, P., Messent, F., 2014, The Role of Public Finance in Deploying Geothermal: Background Paper, Report for Climate Investment Funds.
6. Rybach, L., 2010, Legal and Regulatory Environment Favorable for Geothermal Development Investors, In *Proceedings World Geothermal Congress 2010*, Bali, Indonesia, 7 pp. Available at: www.geothermal-energy.org/pdf/IGAstandard/WGC/2010/0303.pdf.
7. Salmon, J.P., Meurice, J., Wobus, N., Stern, F., Duaime, N., 2011, *Guide to Geothermal Power Finance*, National Renewable Energy Laboratory, France. Available at: www.nrel.gov/docs/fy11osti/49391.pdf.

7

Impact of Geothermal Energy on Climate Change and Tackling It with Machine Learning

7.1 Introduction to Climate Change

As per Sustainable Energy for All (SE4ALL), over 1 billion people lack access to good-quality fuels, and many more, over 2.9 billion, lack energy supplies that allow clean cooking. It is still a huge problem to provide these homes with affordable and dependable energy supplies (Spalding-Fecher et al., 2005). Aside from considerable fuel poverty in several nations, global energy demand is anticipated to rise by 33% until 2040 (IEA, 2015).

To meet expanding energy demands, provide access to billions of people who lack access to advanced fuels, and reduce greenhouse gas (GHG) production, a fundamental shift away from fossil-fuel-based industry scenarios is required. An energy framework that encourages the transformation of our existing power systems to depend on sustainable low-carbon energy supplies should replace the previous emphasis. Sustainable Energy Development (SED) is the name given to this clean energy model, which is described as "the provision of necessary energy services at a reasonable cost in a safe and environmentally friendly fashion, in accordance with economic and social development requirements" (IAEA/IEA, 2001).

Several worldwide efforts to aid SED were established. The SE4ALL program and the Sustainable Development Goals are two examples. The SE4ALL project, which was started in 2011, has three aims: i) Making renewable energy services available to everyone; ii) doubling the global pace of improving energy efficiency; and iii) doubling the worldwide proportion of renewable energy in the energy mix.

Recognizing the significance of SED in achieving global development, the very first SE4ALL Summit in 2014 pledged support for integrating energy in the Sustainable Development Goals (SDGs), which were to substitute the Millennium Development Goals. This approach was successful, and SDG Goal 7 is scheduled to be met by 2030, requiring "secure access to affordable, dependable, sustainable, and affordable energy for everyone".

DOI: 10.1201/9781003204671-7

The United Nations Millennium Declaration was issued at the Millennium Summit in 2000.

The eight Millennium Development Goals (MDGs) were established from the Millennium Declaration and were set to be achieved by 2015.

The MDGs were created with the goal of promoting growth and improving socio-economic circumstances in the poorest nations, diverting the attention of the sustainable development agenda to poverty, civil rights, and the rights of individual populations. The eight MDGs were as follows:

- Goal 1: Ensure that extreme poverty and starvation are no longer a problem.
- Goal 2: Ensure that all children have access to primary education.
- Goal 3: Women should be empowered and gender equality should be promoted.
- Goal 4: Reduce the death rate of children.
- Goal 5: Improve the health of mothers.
- Goal 6: HIV/AIDS, influenza, and other diseases are being combated.
- Goal 7: Ensure long-term environmental viability.
- Goal 8: Create a worldwide development cooperation.

After the Rio+20 meeting in 2012, the 17 SDGs were adopted in 2015 to replace the MDGs (shown in Table 7.1). Being part of a major sustainable development strategy, the SDGs' ultimate goal is to "eliminate poverty, safeguard the planet, and provide prosperity for all". Each goal contains specific objectives that must be met by 2030.

The SDG 7 goal is to "guarantee that everyone has access to adequate, dependable, secure, and modern energy". This goal has three objectives: i) Universally recognized access to quality, dependable, and modern energy supplies; ii) a significant expansion of renewable energy sources in the global energy mix; and iii) doubling the worldwide rates of improving energy efficiency.

Improved access to advanced renewable energy is critical for economic and social growth. In this perspective, the placement of geothermal energy, which is often in isolated rural regions, its capacity to be captured in the smallest form, its reliability, and its ability to shield individual decision makers from climatic and geopolitical events are all essential. These qualities allow remote and rural locations that aren't linked to get access to dependable, high-quality energy (Shortall et al., 2015).

Participation in geothermal energy minimizes import dependency, boosts energy supplies, and raises the proportional share of renewables in total primary energy usage because geothermal is regarded as a renewable source and is produced and consumed locally in most cases.

TABLE 7.1

Sustainable Development Goals Adopted by Millennium Development Goals

Goal 1: Put an end to poverty in all of its manifestations around the world.
Goal 2: Ending hunger, achieving food security and improved nutrition, and promoting sustainable agriculture are all goals that must be met.
Goal 3: Ensure that all people of all ages have good health and are well-adjusted.
Goal 4: Ensure that all students receive a high-quality education that is mutually supportive, and that learning and development opportunities are available to all.
Goal 5: Ensure that all women and girls are empowered through achieving gender equality.
Goal 6: Ensure universal access to clean water, as well as long-term management.
Goal 7: Ensure that everyone has access to health care, dependable, sustainable, and contemporary energy.
Goal 8: Encourage long-term, inclusive, and enhancing economic, as well as full and constructive jobs and decent labour for all people.
Goal 9: Build resiliency in facilities, encourage productivity and competitiveness, and encourage innovation.
Goal 10: Reduce intra- and inter-country inequity.
Goal 11: Make cities and urban development more inclusive, secure, resilient, and long-lasting.
Goal 12: Ensuring that production and trade trends are sustainable.
Goal 13: Take immediate action to address global warming and its consequences.
Goal 14: Oceans, seas, and coastal habitats should be conserved and used responsibly for long-term development.
Goal 15: Preserve, repair, and encourage the ecological use of ecological systems; forest plantations sustainably; battle desertification; and prevent and rectify soil erosion and loss of biodiversity.
Goal 16: To ensure long-term growth, foster world peace, ensure universal legal representation, and construct effective, responsible, and accessible institutions.
Goal 17: Improve and revive the successful community for sustainable development plans implementation mechanisms.

The link between energy usage and population growth and economic development is captured by energy use per gross domestic product (GDP). At capital gains rates, it is desired to lower the amount of energy necessary to create a unit of GDP, for example (Goldemberg and Johansson, 2004). Enhanced geothermal power use can help achieve this aim by increasing "direct use" of thermal energy or by combining energy generation with immediate use of heat energy. Direct use is considerably more efficient than geothermal energy production and imposes less stress on the heat resource's thermal needs. As a result, direct consumption is more cost-effective than thermal power generation in many locations. Cogeneration via a geothermal power plant, or smaller wells or heat pumps such as geothermal energy, can provide "heat" for direct consumption. Warm water from hot springs can be injected directly into the heat pump system, which can then be used for industrial or other economic purposes. Heat transfer, on the other hand, can be used to capture heat in locations where the land is dry but warm. This is also conceivable in "cold" places through the use of heat pumps, which take advantage of the surrounding ecosystem's heat gradient. As a result, it is possible to obtain heat in practically all locations more cost-effectively and efficiently than with traditional furnaces (Lund, 2006).

Although primary geothermal power usage is efficient, secondary geothermal energy uses vary in efficiency depending on the conditions of the geothermal energy and the kind of plant technology utilized.

Generally, geothermal power plants have a poor energy density, varying from 9% to 23%. Unless cogeneration happens and the warm water is used directly and regionally, such as in greenhouses, industrial uses, fisheries, or heat pumps, effluent heat is squandered (Fridleifsson et al., 2008). Therefore, if geothermal energy is being used indirectly for power production, it is critical to guarantee that residual fluids are utilized at low energy concentrations or reinjected (Shortall et al., 2015). From such a brief overview, it is obvious that perhaps the growth of geothermal energy supplies may make a substantial contribution to SDG 7.

7.1.1 Goal 13: Climate Change

Goal 13 encourages nations to take action on climate change as soon as possible. Because greenhouse emissions for each kWh of power obtained from high-enthalpy geothermal fields are much lower than those produced from fossil fuels, geothermal power growth contributes greatly to greenhouse effect mitigation. Emissions of GHGs in units of weight/kilowatt hour range from 4 to 740 g, with a weighted sum of 122 g kWh^{1}, according to Fridleifsson et al. (2008). Fossil fuel power stations emit 740–910 g kWh^{1}, and natural gas combined cycle plants emit 410–650 g kWh^{1} (Schlömer et al., 2014). Geothermal energy production, in terms of having a large impact on abatement, also benefits climate change adaptation (Ogola et al., 2012).

7.2 Mitigation and Geothermal Energy

Geothermal energy technologies have a minimal environmental impact, are unaffected by global warming, and also have the ability to be the world's most affordable type of thermal energy for zero-emission, base-load domestic usage and electricity production. Geothermal power is well situated to play a vital role in clean energy and mitigating climate change measures due to its inherent heat storage capacity, making it ideal both for foundation electric power production and for completely transmissible heating purposes in facilities (Bromley et al., 2010).

Geothermal power has many key attributes that make it ideal for mitigating climate change. These include worldwide reach; native resource; output irrespective of season; immunity to climate and weather environmental change; and effectiveness for on- and off-grid projects as well as base-load electricity production. This foundation energy can also be utilized to refuel rechargeable batteries for automobiles during off-peak hours, reducing carbon emissions by hydrocarbon vehicles. Thermal, geopressure, dry hot rock, and lava are the four major types of geothermal sources (Stober and Bucher, 2013). When magma rises close enough to the top to heat water sources confined in cracked or porous formations, or when water flows at greater depth across faults, hydrothermal deposits emerge (DiPippo, 2008; Stober and Bucher, 2013; Xydis et al., 2013). Based on conditions of reservoir temperature, ability to produce electricity, and other factors, they can be employed for a variety of energy uses. Low-enthalpy thermal sources, which are typically around 100 °C, could be utilized directly for balneology, to warm buildings, to produce food, for aquaculture, and for other purposes, and are therefore not ideal for production of electricity (Allen and Milenic, 2003; Xydis et al., 2013). This could be accomplished using either a closed or open circuit, depending on the underground system's features (Preene and Younger, 2014). Groundwater doesn't really circulate in a closed-loop system, instead exchanging energy with a vertical or horizontal circuit of water coolant mixture; in an open-loop circuit, groundwater serves as a chiller to transmit thermodynamic heat (Lund et al., 2004). Geothermal energy with temperatures above 100 °C can be utilized for electricity production (DiPippo, 2008; Stober and Bucher, 2013). Vapour-dominated dry steam and liquid-dominated hot water reservoirs are the two main forms of geothermal resources for producing power (DiPippo, 2008; Stober and Bucher, 2013). Dry power stations use subterranean steam to generate electricity. To drive a turbine or generator section, steam is pumped straight from subterranean boreholes to the facility (DiPippo, 2008). The most frequent type of flash steam cycle is one that uses geothermal reserves of fluid with temperatures above 182 °C. By its own pressure, the warm water rises via wells in the subsurface. As it rises, pressure drops, and heated water is converted into steam, which is used to generate electricity. Any remaining water and condensed steam are pumped back into reservoir. Since water in a thermal reservoir isn't hot enough just to turn into vapour, binary cycle plants are utilized, and they run on water from 107 ° to 182 °C (Stober and Bucher, 2013). Such facilities heat the fluid to drive the turbine using organic materials with a low vapour pressure; water is injected directly underground to be warmed (DiPippo, 2008; Stober and Bucher, 2013).

To mitigate the environmental effects of geothermal activities, a variety of strategies could be used. Detailed monitoring during all phases of exploration, construction, operation, and decommissioning will provide a preventive approach and the opportunity to address negative impacts before they become too challenging (Table 7.2).

Since it supplies heating from a large, scattered electrical source, geothermal energy generation is best adapted to reducing climate change. It has a strong track record of producing expensive and extremely low base power using current technology, which is suitable for both industrial and developing nations. Space heating systems, as well as a number of many other commercial processes, are all possible with geothermal energy-heated fluids, and range from minor to district-wide systems. Furthermore, ground source heat pumps (GHPs) are being used all over the world, allowing significant improvements in house heating and cooling effectiveness. As contrasted with certain other renewables, heating assets have a comparatively high availability index. EGS (enhanced or engineered geothermal

TABLE 7.2

Mitigation of Geothermal Energy

Phase of Impact	Mitigation Strategies That Could Be Used
Plant development preparation	The power facility might be enclosed to keep wildlife out, avoid ecologically vulnerable areas, and preserve natural and historical aesthetic sites.
During the design and renovation of the plant and machinery	Follow procedures for induced and natural seismic events in heating systems. This may differ per nation. Once project is performed, reduce the risk of surface disturbance by landscaping. If neighbourhoods are being impacted, put up noise barriers.
During the construction and operation of the facility	The facility must be built in such a way that no steam escapes into the environment. At the cooling system, non-condensable pollutants are treated or diluted with air. Detect H_2S pollutants from flash geothermal power plants to see if they are under pollution controls and act accordingly. To prevent gas accumulation in restricted spaces, use ventilation. Controlling water spillage from geothermal facilities into the land and nearby below- and above-ground surface waters is also a priority. Instead of using inhibition, use thermodynamic levelling means to prevent polluted solid and liquid waste from spilling. To reduce the possibility of erosion and extend the reservoir's duration, include reinjection of water into reservoir maintenance from the start.
Following decommissioning	Consider following rehabilitation techniques for disposing of toxins, sweeping up equipment, and returning damaged lands to resistant, ecological functions that mirror local ecosystems.

(Dhar et al., 2020)

systems) have the ability to be used on a worldwide basis whenever or wherever they are required. To realize this promise, industrial-scale data and experimental testing are necessary, as well as multi-year state and corporate assistance and funding. Global warming prevention will be achieved by the distribution of CO_2 credits at affordable prices, as well as increased energy supplies. In numerous nations, like Costa Rica (19%), the Philippines (19%), El Salvador (24%), Kenya (19%), and Iceland (30%), geothermal power accounts for a significant portion of overall electricity consumption (15%). In more than 70 nations, thermal power can be used for direct heat applications such as room heating; greenhouses and fishery pool heating; food drying; chemical products; balneology; refrigeration; melting snow; and GHPs. The possible effects of global warming as well as other harmful ecological repercussions of fossil-fuelled electricity and heat production are currently one of the major thermal expansion opportunities. This has increased public awareness of the importance of using sustainable energy sources like thermal. In the International Energy Agency's (IEA) reference period, thermal power generation is expected to rise to 350 TWh by 2050 and to 1060 TWh later. In 2050 and well beyond, even more thermal output possibilities can be realized with a more pro-active attitude toward increasing research and installation. Preliminary capital costs, an insufficient knowledge of geothermal power, ambiguity of government incentives, and imagined environmental issues are all major impediments to growth. Despite the fact that GHP devices provide thermal energy, they require a power source. In existing methods, the coefficient of performance (COP) and the proportion of extraction to usage are around 3 and 5, respectively. This indicates that even a GHP system can save around 66% and 80% of its actual extraction capacity in terms of power used in warming buildings. The GHP technique offers great ability to significantly cut carbon emissions and hence can be pushed by governments to minimize future CO_2 emissions, particularly in rural communities that currently rely on fossil fuels for electricity production. A considerable portion of worldwide renewable power is consumed to supplant fossil fuel production and so alleviate the consequences of climate change from carbon dioxide emissions.

7.3 Factors Influencing Adaptation and Mitigation

The United Nations Framework Convention on Climate Change (UNFCCC) distinguishes between two types of climate change responses: mitigation, which involves lowering GHGs and increasing sinks, and adaptability, which involves adjusting to the effects of climate change. As participants to the UNFCCC as well as the Paris Agreement, the majority of industrialized countries have committed to developing government policies and implementing associated actions to mitigate climate change and reduce their net carbon output (United Nations, 2020). Prevention (cutting emissions) and adaptability are two aspects of dealing with climate change. Both of these challenges are intricate. Given a better knowledge of climate and catastrophic occurrences, adaptation necessitates catastrophe preparation and disaster response. Modifications in electrical networks, mobility, buildings, manufacturing, and land use are all required to reduce GHG production.

The Kyoto Pact established stringent standards to cut emissions by 5.2% relative to 1990 during the first budgetary term of the Protocol, which ran from 2008 to 2012. On 16 February 2005, the Agreement went into effect, making the Protocol's reductions objectives legally enforceable for countries that had signed it. The Kyoto Protocol was built around market-based measures for reducing GHG emissions. Versatility measures were developed to expand the range of national options for reducing GHGs; lowering costs and development were key for low-carbon alternatives in nations that had not committed to reduction.

The different components of extensibility were (UNFCCC, 2016):

1. **Global Emission Market,** which enabled a nation to augment its carbon credit to nations whose pollution exceeded their limits if they lowered pollution under their objectives. The EU Emissions Trading Framework comprises such a world trade organization and thus a coal exchange (EU-ETS).
2. **Joint Implementation (JI)**, which enabled nations with enforceable emission targets to receive reducing emissions units (ERUs) by providing a small amount of renewable energy in many other nations with abatement obligations.
3. **The Clean Development Mechanism (CDM)**, which allows nations with enforceable reducing emissions commitments to gain certified energy efficiency credits by investing in developing-country programmes that result in "additional" emissions reduction.

Protection of the environment was also a goal of the Paris Agreement, and financing mechanisms for adaptability were developed through the Right Provider. The Adaptation Financing was funded in the very first financial term of the Kyoto Protocol by a part of the profits from CDM operations. Later, in 2012, it was recommended to extend the budget of the Right Provider to also include 2% of JI and global carbon trading earnings (UNFCCC, 2016). Adaptation is "Alteration in internal or external change in relation to existing or anticipated climatic stressors or their impacts, which mitigates harm or exploits good opportunities," according to the IPCC. Instinctive and reactionary adaptability, personal and communal adaptation, and independent and organized adaptation are all examples of different types of adaptation. Resilience, as per the IPCC, comprises both activities that mitigate harm and measures that permit reaping the benefits of global warming or global warming mitigation possibilities: "Caused by human activity action to decrease the production or boost the concentrations of greenhouse gas emissions," according to the IPCC. Stabilization aims to avoid effects of climate change GHGs by man-made initiatives and increasing sequestration, such as in forests as well as other plants.

Adopting interconnected wastewater treatment, enhancing sources, lowering the discrepancy in water availability, lowering non-climate stress factors, reinforcing infrastructure facilities, and trying to adopt the most water-efficient innovations and liquid strategies are some of the responsive water management techniques and strategies.

Due to increased implementation of global surveillance, modelling, and resource evaluations, fisheries as well as some fish farming businesses with high-technology and/or big expenditures have considerable capacity for adaptability. Adaptation options include large-scale translocation of industrial fishing activities and flexible management that can react to variability and change. For smaller-scale fisheries and nations with limited adaptive capacities, building social resilience, alternative livelihoods, and occupational flexibility are important strategies. Adapting methods for coral reef areas are typically confined to decreasing other stresses, mostly by water purification and minimizing demands from fishing and tourism, but as extreme temperatures and ocean acidification grow, their effectiveness will be significantly decreased. Agricultural adaptation alternatives include producing new varieties of crops adaptable to increases in CO_2, heating, and drought, strengthening the capacity for managing environmental risk, and mitigating the financial ramifications of deforestation. Enhancing financial assistance and participating in small-scale agricultural productivity can also be beneficial. Remediation could be more cost-effective when combined with initiatives to minimize energy consumption and GHG concentration in end-use industries, deal with climate change energy supply, decrease carbon emission, and improve carbon sinks in land-based industries. Techniques that establish a price for carbon, such as carbon trading schemes and carbon pricing, can accomplish cost-effective abatement in theory, but they have been implemented with varying results due to country conditions as well as policy development.

Figure 7.1 shows the factors that influence climate change mitigation and adaptation.

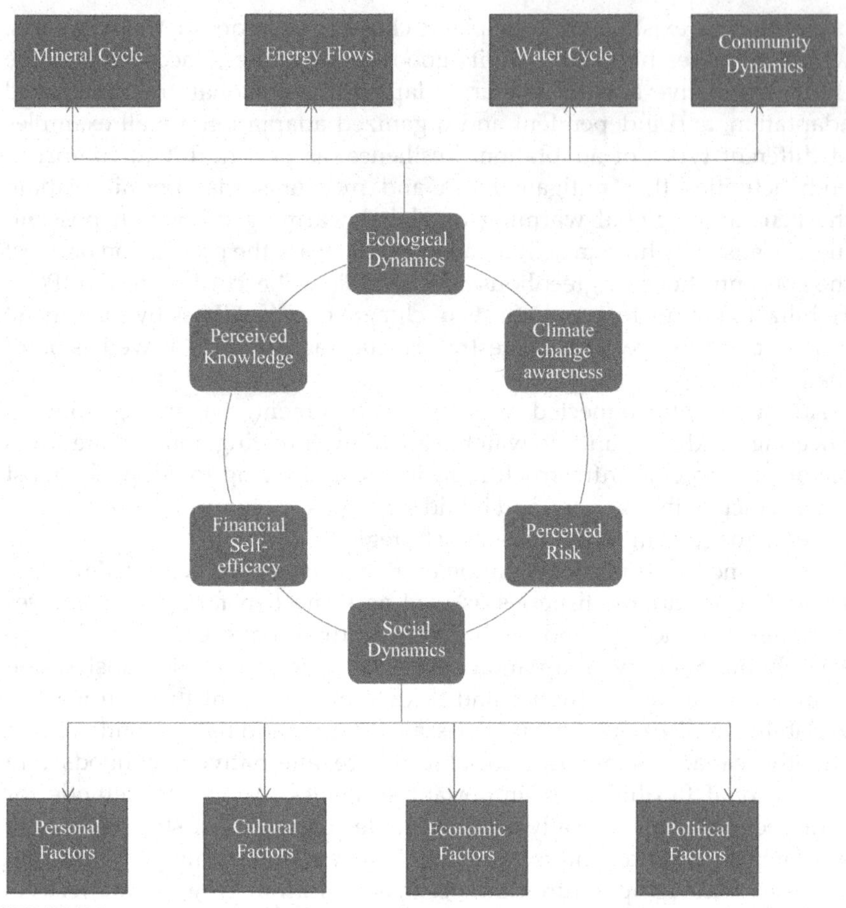

FIGURE 7.1
Factors influencing climate change mitigation and adaptation.

Frame interpolation of low- and zero-carbon power production methods is required for carbon reduction of the energy supply industry, that is, lowering the carbon emissions. Replacement of present global coal-fired power stations with advanced, extremely effective natural gas combined cycle or heating and power stations can considerably reduce GHG emissions from energy production, provided natural gas is accessible and fugitive emissions linked with harvesting and stockpiling are low or remedied. Planting trees, forest resources administration, and limiting deforestation are by far the most cost-effective forestry abatement methods, with considerable changes in their relative relevance between regions. Agricultural land administration, grazed forest management, and soil organic regeneration are by far the most cost-effective mitigation measures in agriculture. Biomass has the potential to play a significant part in climate change mitigation, but there are several challenges to consider, such as the long-term viability of techniques and the effectiveness of biofuel infrastructure. Research demonstrates that biomass solutions with minimal life cycle emissions can contribute to minimizing GHGs; however, the results are geographically limited and depend on integrated transport "biomass to bioenergy networks" as well as sustainable land usage governance and management. Concerns regarding GHGs from land, nutrition security, water management, environmental protection, and lifestyles are all obstacles to huge bioenergy implementation.

7.4 Application of Machine Learning in Climate Change Mitigation

In the year 1896, Arrhenius predicted that carbon pollution would eventually produce enough CO_2 to raise the world temperature by 5 °C. The projections have become significantly more detailed and exact, despite the fact that the federal mandates supporting those computations have not altered. Climate change models, also known as General Circulation Models (GCMs) or Earth System Models (ESMs), are the most frequently used prediction tools. These models enable people to quantify their climate change risks and anticipate their possible implications, which helps countries in the region to make better decisions. According to a 2018 UN assessment on changing climate, unless global carbon emissions are stopped within 30 years, the planet will face disastrous repercussions. Machine learning (ML) has gained popularity in recent years as a versatile tool for technological advancement. Despite the rise of movements using ML and artificial intelligence (AI) to solve societal and global problems, there is still a need for a thorough investigation of how these technologies might best be used to combat climate change. AI can be used to make a significant difference in the battle against climate change, whether through efficient engineering or cutting-edge studies. AI and ML are becoming more prevalent in people's lives thanks to considerable improvements in processor supply, speed, connection, and low-cost data management. Data analytics, AI, and ML are just a few of the analytical models available. Data analytics is focused on the use of large datasets that are too huge to be analysed using conventional methods. AI is a branch of computer science whose purpose is to train a device to perform jobs that a human cannot, and which typically involves judgment in a variety of situations. ML is a branch of AI in which computers discover connections from massive classification algorithms. A simple characterization can be used for climate and weather purposes, such as: (i) Data analysis is defined as the collecting of meteorology or Earth System-related observations, as well as high time- and frequency-resolution ESM results, for evaluation; (ii) Improving or exploring different connections across regions, times, and variables in databases (e.g. where ocean temperature characteristics improved weather forecast months ago over areas of the country) are examples of ML; and (iii) AI is a means of providing automatic warnings and guidance to society in the event of oncoming weather extremes, based on the links discovered by ML. The usage of computer graphics processing units (GPUs) has recently made it easier to apply ML approaches due to increased computational capabilities. In general, there are two types of ML methodologies: supervised learning and unsupervised (Murphy, 2012). i) Supervised approaches transfer feeds to output results and depend on an advanced definition of dependent variables. Explanatory factors, such as massive forces like GHG levels, teleconnection variables, or measurements of a specific area of the earth's surface, are examples of feeds. Outcomes are interesting guiding principles, such as local climatic effects. Supervised techniques make use of a training sample set that includes both observed inputs and outputs. ii) Only the outcomes of gathered data are used in unsupervised techniques, and the goal is to find interesting patterns and relationships as well as relationships to inputs that aren't mentioned up front. Unsupervised learning has the potential to aid in the production of new linkages, or teleconnections, between distinct aspects of climate modelling. ML methods have been used by climate scientists to increase their understanding of the various major earth systems. Table 7.3 shows how ML can be used in climate science.

7.4.1 Gradient Descent Method

Gradient descent is similar to normal regression analysis in that it seeks to minimize the variance ("cost function", F), which is expressed as Eq. (7.1)

$$U\ (n+1) = U\ (n) + \gamma\ \mathrm{d}F\ (U\ (n)) /\ \mathrm{d}U\ (n) \tag{7.1}$$

U(n) = is a discrete integer (or vector) and is a threshold value that ensures that the objective function gets smaller with each repetition (Eq. (7.2)).

$$\mathrm{F(U(n)) > F(U(n+1))} \tag{7.2}$$

Where n = number of iterations.

The pace of the gradient descent strategy γ is determined by a variable that can be regarded as a training variable (or a time step). Ineffective algorithms result from poor optimization of γ.

TABLE 7.3

Existing Application of Machine Learning Algorithms to Climate Science

Component of Climate Change Research	Technique Used with the Standard Acronym	Finding or Development	References
Future climate scenarios	Bayesian linear regression	By weighing models by skill, you can combine various climate patterns projections.	Luo et al. (2007)
Future climate scenarios	Generalized Hidden Markov Models (HMM)	Climate models that are weighted according to their skill values greater than ensemble standards.	Monteleoni et al. (2011)
Climate impacts	Support Vector Machines (SVM), Artificial Neural Networks (ANN), Generalized Regression Neural Network (GRNN).	Examine how climate change will affect above-ground bioenergy.	Wu et al. (2019)
Climate impacts	Model Tree Ensemble (MTE)	Examine the effects of climate change on the worldwide water cycle, with a focus on evaporation alterations.	Jung (2010)
Climate impacts	Principal Components Analysis (PCA) and fuzzy clustering. Relevance Vector Machine (RVM).	Evaluate the influence of impacts of climate change on India's hydrological systems, particularly river flow.	Ghosh and Mujumdar (2008)
Climate impacts	Fuzzy logic, Least Squares Support Vector Regression (LS-SVR), Artificial Neural Networks (ANN), Adaptive Neuro-Fuzzy Inference System (ANFIS)	In India, forecast hydrological factors (evaporation) based on meteorological factors (rain, humidity).	Goyal et al. (2014)
Climate impacts	Model Tree Ensembles (Random Forests, RF)	Analyse how different environmental systems are affected by water shortage (drought).	Yang et al. (2016)
Climate impacts	Convolutional Neural Network (CNN), Gaussian Process (GP) Regression	Crop yields can be estimated using satellite imagery.	Azzari et al. (2017), Burke and Lobell (2017)
Climate impacts	Artificial Neural Networks (ANN)	Determine how climate variables affect sand accumulation in semi-arid areas.	Buckland et al. (2019)

(*continued*)

TABLE 7.3 (Continued)

Existing Application of Machine Learning Algorithms to Climate Science

Component of Climate Change Research	Technique Used with the Standard Acronym	Finding or Development	References
Climate datasets	Random Forest (RF)	To examine hydrological patterns and volatility, create a long-term, consistent global runoff database.	Ghiggi et al. (2019)
Climate datasets	Artificial Neural Network (ANN)	To produce globally reasonable estimate of precipitation, use satellite-based extraction (PERSIANN).	Hsu et al. (1997), Hong et al. (2007), Nguyen et al. (2018)
Climate datasets	Gaussian Process (GP) model fitted with a Markov Chain Monte Carlo (MCMC) method	For inadequate data series, enhancing predictions of minimum and maximum temperature. Improve maximum recommended and lowest temperature estimations based on data from many other adjacent observations.	Rischard et al. (2018)
Climate datasets	Kernel Regression (KR)	GCM meteorological fields should be downscaled to suitable scales for effect evaluation.	Salvi et al. (2017)
Climate extremes	Convolutional Neural Network (CNN); 3D Convolutional encoder-decoder	Detect extreme climate occurrences in a climate change model's results.	
Climate extremes	Extreme Learning Machine and Convolutional Neural Network (CNN)	Utilizing atmospheric and climate variables as inputs, forecast a drought rating.	Deo and Sahin (2015)
Climate extremes	Random Forest (RF); Gradient Boosted Regression Trees (GBRT)	Using satellite data, forecast environmental and agronomic drought situations.	Park et al. (2016)
Climate extremes	Artificial Neural Network (ANN); Support Vector Regression (SVR); Wavelet Transforms	In Ethiopia, previous climatic data is used to predict meteorological catastrophes.	Mishra and Desai (2006), Belayneh et al. (2016)

7.4.2 Gaussian Processes

Gaussian dynamics (Rasmussen and Williams, 2006) is a supervised ML method that can be used instead of linear regression methods. A Gaussian dynamics is a bunch of different factors, Y (e.g. data observations), with a multimodal confidence interval for any set of these factors. Instead of input state, x, the Gaussian model is defined over measurement function, Y. A median function and a correlation coefficient outline the scope.

7.4.3 Non-linear, Non-Gaussian Inferences

The objective is to parameterize hidden Markov models using ML for forecasting and reasoning from spatio-temporal phenomena. State-space difficulties exist in climate change science, in which the processes are implicit or "hidden" and only measurements on alternative data can be collected. The most straightforward problem is a Kalman filter (Eqs. (7.3) and (7.4)):

$$Y(t) = A(t) + \epsilon_1 \tag{7.3}$$

$$A(t) = A(t-1) + \epsilon_{21} \tag{7.4}$$

Where
Y(t) = is observation at time t (with Gaussian error ϵ_1)
A(t) = latent state.

A is a static process that needs estimate, such as immutable input parameters in the global climate (Schneider et al., 2017). Several state-space issues are complex and non-Gaussian, necessitating Bayesian network approaches.

7.4.4 Deep Learning Approaches

A prototype system is used in deep learning algorithms. Data is entered at the bottom of the graph, altered via hidden layers, and released at the top. Masses are linked with vertices in networks, and, known as "biases", are connected with neurons (nodes), in which the masses determine the characteristics of connectivity among neurons from multiple layers, as well as producing an imbalance that controls the neuron's responsiveness. Using testing data, these course outlines can provide out-of-sample forecasts. The capacity of a system to capture characteristics of recent anomalies (i.e. "out of sample" events) enhances confidence in projections of future abnormalities, which could become frequent, similarly to Gaussian dynamics. By giving fresh perspectives into the immensely rich range of interrelated Earth System behaviours and their many connections with biological cycles, ML implementation to the earth's surface will be in the effective class. For environmental assessment, AI is well established. While ML will expose climate system features and improve predictions across timeframes, AI will be able to use this data to make judgments. Where ML becomes AI is in the teaching of activities required to guarantee safety throughout environmental extremes. ML adoption will likely facilitate the stage process required to produce enhanced advice on the climatic conditions envisaged for increased GHG emissions in overall environmental policy. Deep learning models have recently been merged with current thermodynamic understanding to correct the main range of risk in current environment models: Clouds, according to new research. Storm clouds absorb outbound heat to maintain the earth's temperature, whereas bright skies block sunlight and cool the earth. Ozone levels may be calculated not as a function of time and temperature but instead by class distinction rules, saving time and money. New ML models for environment will have the best chance of succeeding if they are tightly linked with established climate theories.

References

1. Allen, A., Milenic, M., 2003, Low-enthalpy Geothermal Energy Resources from Groundwater in Fluvioglacial Gravels of Buried Valleys, *Applied Energy*, Vol. 74, pp. 9–19. Available at: https://doi.org/10.1016/S0306-2619(02)00126-5.
2. Azzari, G., Jain, M., Lobell, D.B., 2017, Towards Fine Resolution Global Maps of Crop Yields: Testing Multiple Methods and Satellites in Three Countries, *Remote Sensing of Environment*, Vol. 202, pp. 129–141.
3. Belayneh, A., Adamowski, J., Khalil, B., Quilty, J., 2016, Coupling Machine Learning Methods with Wavelet Transforms and the Bootstrap and Boosting Ensemble Approaches for Drought Prediction, *Atmospheric Research*, Vol. 172, pp. 37–47.
4. Bromley, C.J., Mongillo, M., Hiriart, G., Goldstein, B., Bertani, R., Huenges, E., Ragnarsson, A., Tester, J., Murako, H., Zui, V., 2010, Contribution of Geothermal Energy to Climate Change Mitigation: The IPCC Renewable Energy Report, *Proceedings World Geothermal Congress 2010*, Bali, Indonesia, 25–29 April 2010.
5. Buckland, C.E., Bailey, R.M., Thomas, D.S.G., 2019, Using Artificial Neural Networks to Predict Future Dryland Responses to Human and Climate Disturbances, *Scientific Reports*, Vol. 9, p. 3855.
6. Burke, M., Lobell, D.B., 2017, Satellite-based Assessment of Yield Variation and Its Determinants in Smallholder African Systems, *Proceedings of the*

National Academy of Sciences of the United States of America, Vol. 114, pp. 2189–2194.

7. Deo, R.C., Sahin, M., 2015, Application of the Extreme Learning Machine Algorithm for the Prediction of Monthly Effective Drought Index in Eastern Australia, *Atmospheric Research*, Vol. 153, pp. 512–525.
8. Dhar, A., Naeth, M.A., Jennings, P.D., Gamal El-Din, M., 2020, Geothermal Energy Resources: Potential Environmental Impact and Land Reclamation, *Environmental Reviews*, Vol. 28 (4), pp. 415–427.
9. DiPippo, R., 2008, *Geothermal Power Plants: Principles, Applications, Case Studies, and Environmental Impact*, 2nd ed., Butterworth-Heinemann, Elsevier, UK.
10. Fridleifsson, I.B., Bertani, R., Huenges, E., Lund, J.W., Ragnarsson, Á., Rybach, L., 2008, The Possible Role and Contribution of Geothermal Energy to the Mitigation of Climate Change, In *IPCC Scoping Meeting on Renewable Energy Sources – Proceedings. Intergovernmental Panel on Climate Change*, edited by O. Hohmeyer and T. Trittin, Elsevier, Lübeck, Germany, pp. 59–80.
11. Ghiggi, G., Humphrey, V., Seneviratne, S.I., Gudmundsson, L., 2019, GRUN: An Observations-based Global Gridded Runoff Dataset from 1902 to 2014, *Earth System Science Data Discussions*, Vol. 11, pp. 1655–1674.
12. Ghosh, S., Mujumdar, P.P., 2008, Statistical Downscaling of GCM Simulations to Streamflow Using Relevance Vector Machine, *Advances in Water Resources*, Vol. 31, pp. 132–146.
13. Goldemberg, J., Johansson, T.B. (editors), 2004, *World Energy Assessment Overview: 2004 Update*, United Nations Development Programme, New York, USA, prepared for United Nations Development Programme, United Nations Department of Economic and Social Affairs, and the World Energy Council, 88 pp.
14. Goyal, M.K., Bharti, B., Quilty, J., Adamowski, J., Pandey, A., 2014, Modeling of Daily Pan Evaporation in Sub Tropical Climates Using ANN, LS-SVR, Fuzzy Logic, and ANFIS, *Expert Systems with Applications*, Vol. 41, pp. 5267–5276.
15. Hong, Y., Gochis, D., Cheng, J-T., Hsu, K-L., Sorooshian, S., 2007, Evaluation of PERSIANN-CCS Rainfall Measurement Using the NAME Event Rain Gauge Network, *Journal of Hydrometeorology*, Vol. 8, pp. 469–482.
16. Hsu, K.L., Gao, X.G., Sorooshian, S., Gupta, H.V., 1997, Precipitation Estimation from Remotely Sensed Information Using Artificial Neural Networks, *Journal of Applied Meteorology*, Vol. 36, pp. 1176–1190.
17. International Atomic Energy Agency (IAEA) / International Energy Agency (IEA), 2001, Indicators for Sustainable Energy Development. Paper presented at 9th Session of the United Nations Commission on Sustainable Development, New York, USA, 20 pp.
18. International Energy Agency (IEA), 2015, *Energy Technology Perspectives 2015*, Organisation for Economic Co-operation and Development (OECD) and International Energy Agency (IEA), Paris, France, 418 pp.
19. Jung, M., 2010, Recent Decline in the Global Land Evapotranspiration Trend due to Limited Moisture Supply, *Nature*, Vol. 467, pp. 951–954.

20. Lund, J., Sanner, B., Ryback, L., Curtis, R., Hellstrom, G., 2004, Geothermal (Ground-Source) Heat Pumps – A World Overview, *Geo-Heat Center Quarterly Bulletin*, Vol. 25 (3), pp. 1–10, Klamath Falls, Oregon.
21. Lund, J.W., 2006, Geothermal Energy Focus: Tapping the Earth's Natural Heat, *Refocus*, Vol. 7–6, pp. 48–51.
22. Luo, L.F., Wood, E.F., Pan, M., 2007, Bayesian Merging of Multiple Climate Model Forecasts for Seasonal Hydrological Predictions, *Journal of Geophysical Research: Atmospheres*, Vol. 112, p. 13.
23. Mishra, A.K., Desai, V.R., 2006, Drought Forecasting Using Feedforward Recursive Neural Network, *Ecological Modelling*, Vol. 198, pp. 127–138.
24. Monteleoni, C. Schmidt, G.A, Saroha, S., Asplund, E., 2011, Tracking Climate Models, *Statistical Analysis and Data Mining: The ASA Data Science Journal*, Vol. 4, pp. 372–392.
25. Murphy, K.P., 2012, *Machine Learning: A Probabilistic Perspective*, MIT Press, Cambridge, MA, p. 1096.
26. Nguyen, P., et al., 2018, The PERSIANN Family of Global Satellite Precipitation Data: a Review and Evaluation of Products, *Hydrology and Earth System Sciences*, Vol. 22, pp. 5801–5816.
27. Ogola, P., Davídsdóttir, B., Fridleifsson, I.B., 2012, Opportunities for Adaptation–Mitigation Synergies in Geothermal Energy Utilization – Initial Conceptual Frameworks, *Mitigation and Adaptation Strategies for Global Change*, Vol. 17–5, pp. 507–536.
28. Park, S., Im, J., Jang, E., Rhee, J., 2016, Drought Assessment and Monitoring through Blending of Multi-sensor Indices Using Machine Learning Approaches for Different Climate Regions, *Agricultural and Forest Meteorology*, Vol. 216, pp. 157–169.
29. Preene, M., Younger, P.L., 2014, Can You Take the Heat? – Geothermal Energy in Mining, *Mining Technology*, Vol. 123 (2), pp. 107–118.
30. Rasmussen, C.E., Williams, C.K.I., 2006, *Gaussian Processes for Machine Learning (Adaptive Computation and Machine Learning*, MIT Press, Cambridge, MA, p. 272.
31. Rischard, M., McKinnon, K.A., Pillai, N., 2018, Bias Correction in Daily Maximum and Minimum Temperature Measurements through Gaussian process modeling, *Environmental Research Letters*, Vol. 14 (2019), p. 124007. arXiv:1805.10214v2.
32. Salvi, K., Villarini, G, Vecchi, G.A., 2017, High Resolution Decadal Precipitation Predictions over the Continental United States for Impacts Assessment, *Journal of Hydrology*, Vol. 553, pp. 559–573.
33. Schlömer, S., Bruckner, T., Fulton, L., Hertwich, E., McKinnon, A., Perczyk, D., Roy, J., Schaeffer, R., Sims, R., Smith, P., Wiser, R., 2014: Annex III: Technology-specific Cost and Performance Parameters, In *Climate Change 2014: Mitigation of Climate Change. Contribution of Working Group III to the Fifth Assessment Report of the Intergovernmental Panel on Climate Change*. Cambridge University Press, Cambridge, UK and New York, NY, USA, 1454 pp.
34. Schneider, T., Lan, S.W., Stuart, A., Teixeira, J., 2017, Earth System Modeling 2.0: A Blueprint for Models That Learn from Observations and Targeted

High-resolution Simulations, *Geophysical Research Letters,* Vol 44, pp. 12396–12417.
35. Shortall, R., Davídsdóttir, B., Axelsson, G., 2015, Geothermal Energy for Sustainable Development: A Review of Sustainability Impacts and Assessment Frameworks, *Renewable and Sustainable Energy Reviews,* Vol. 44, pp. 391–406.
36. Spalding-Fecher, R., Winkler, H., Mwakasonda, S., 2005, Energy and the World Summit on Sustainable Development: What Next? *Energy Policy,* Vol. 33–1, pp. 99–112.
37. Stober, I., Bucher, K., 2013, *Geothermal Energy: From Theoretical Models to Exploration and Development,* Springer-Verlag, Berlin, Heidelberg, Germany. https://doi.org/10.1007/978-3-642-13352-7.
38. UNFCCC, 2016, *Kyoto Protocol. United Nations Framework Convention on Climate Change.* Available at: http://unfccc.int/kyoto_protocol/items/2830.php.
39. United Nations, 2020, *United Nations Climate Change Annual Report, 2020,* ISBN: 978-92-9219-197-9.
40. Wu, C., Chen, Y., Peng, C., Li, Z., Hong, X., 2019, Modeling and Estimating Aboveground Biomass of *Dacrydium pierrei* in China Using Machine Learning with Climate Change, *Journal of Environmental Management,* Vol. 234, pp. 167–179.
41. Xydis, G.A., Nanaki, E.A., Koroneos, C.J., 2013, Low-enthalpy Geothermal Resources for Electricity Production: A Demand-side Management Study for Intelligent Communities, *Energy Policy,* Vol. 62, pp. 118–123.
42. Yang, Y.T., et al., 2016, Contrasting Responses of Water Use Efficiency to Drought across Global Terrestrial Ecosystems, *Scientific Reports,* Vol. 6, p. 8.

Index